CONTENTS

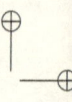

2   CONTENTS

CONTENTS 3

4 CONTENTS

Deltoic "schematic" paradigm

Kliment Sandjakoski, MD

December 19, 2019

Part I
S-orbitalic chronometer

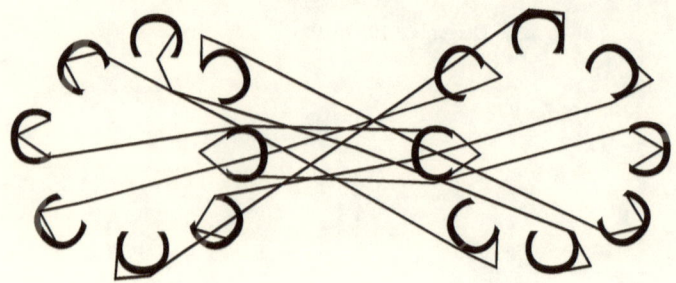

1 Chronome

$$\int_{\circlearrowleft}^{C} \frac{sin\gamma}{cos\gamma} dt dg = T \tag{1}$$

Part II
π–vector alphanumerical matrix

Supine-positioning the cytosine(C) and adenine(A) in terms of distributive π–crossings.

^{1}C	^{2}C	^{3}C	^{4}C	A	^{5}C
^{6}C	A	$^{7}C^{1}$	^{8}C	A	^{9}C
^{10}C	^{11}C	A	A	$^{12}C^{2}$	^{13}C
$^{1}C^{3}$	^{2}C	A	$^{3}C^{4}$	A	$^{4}C^{5}$
A	A	^{5}C	A	A	A
$^{6}C^{6}$	$^{7}C^{7}$	$^{8}C^{8}$	$^{9}C^{9}$	A	$^{10}C^{10}$

0.1 Alternative singularity crossroads

1-1	CCCCA		1→ 10	CCCCAC
2-2	CCCA		2→ 9	CCAC
3-3	CA		3→ 8	CACC
4-4	CCA		4→ 7	CACCC
5-5	CCCCA		5→ 6	CAC
6-6	CCCCCA			
8-8	CCCCA			
9-9	CCCCCA			
10-10	CCCA			
11-11	CA			
12-12	CCCA			
13-13	CCCCCCA			

0.2 Transcriptomal&Translational initiation basics

CCCCCACCCCAC

$\frac{C}{G}\frac{C}{G}\frac{C}{G}\frac{A}{U}\frac{C}{G}\frac{A}{U}\frac{C}{G}\frac{C}{G}$

CCCACACC

$\frac{C}{G}\frac{C}{G}\frac{C}{G}\frac{A}{U}\frac{C}{G}\frac{A}{U}\frac{C}{G}\frac{C}{G}$

CCCCCCACACCC

$\frac{C}{G}\frac{C}{G}\frac{C}{G}\frac{C}{G}\frac{C}{G}\frac{A}{U}\frac{C}{G}\frac{A}{U}\frac{C}{G}\frac{C}{G}$

$\frac{C}{G}\frac{A}{U}\frac{C}{G}\frac{C}{G}\frac{A}{U}\frac{C}{G}$

$\frac{C}{G}\frac{A}{U}\frac{C}{G}\frac{C}{G}\frac{C}{G}\frac{C}{G}\frac{A}{U}\frac{C}{G}\frac{C}{G}\frac{C}{G}$

$\frac{C}{G}\frac{C}{G}\frac{C}{G}\frac{A}{U}\frac{C}{G}\frac{A}{U}\frac{C}{G}\frac{C}{G}$

$\frac{C}{G}\frac{C}{G}\frac{C}{G}A \qquad \frac{}{G}$

$\frac{C}{G}\frac{C}{G}\frac{C}{G}\frac{A}{U}\frac{C}{G}\frac{A}{U}\frac{C}{G}\frac{C}{G}$

$\frac{C}{G}\frac{C}{G}\frac{C}{G}\frac{A}{U}\frac{C}{G}\frac{C}{G}\frac{C}{G}\frac{C}{G}\frac{A}{U}\frac{C}{G}$

AGU codes for the starting aminoacid in transcriptional principle

Deciphering forward the periodic structural approach, configures molecular pathways, shedding light onto basic mechanisms, how genetic aparatus works.

1 Δ-determined secondary structure

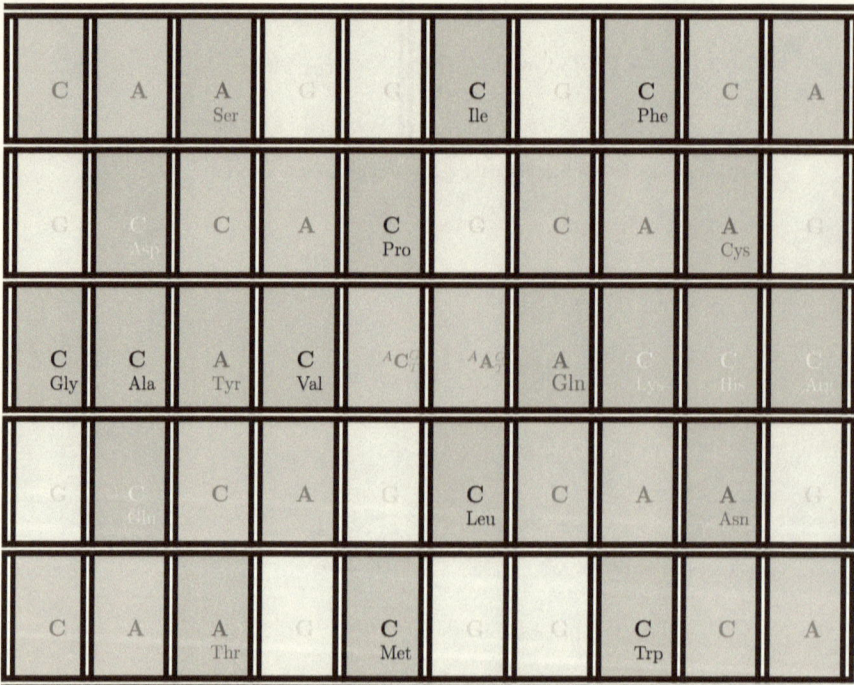

Note the "dollar sign" following aminoacid distribution on a frontal plane

1.1 "Clock to eternity"

The idea is to divide imaginative chart to 4(four) quadrants, and further on to 2(two) subquadrants–8 in total subquadrants representing onw of the four nucleotides, equally distributed.

In addition to the previous step, using the existing table of the genetic code, I projected three arrows(seconds, minutes and hours) by the encoding genetic triplet, accordingly.

The result, presented in the mode of "Pendulum-type of wall clock" is presented below. Seconds-arrow is projected in pendulum style.[1]

Illustrations below demonstrate alternative projections of existing model.

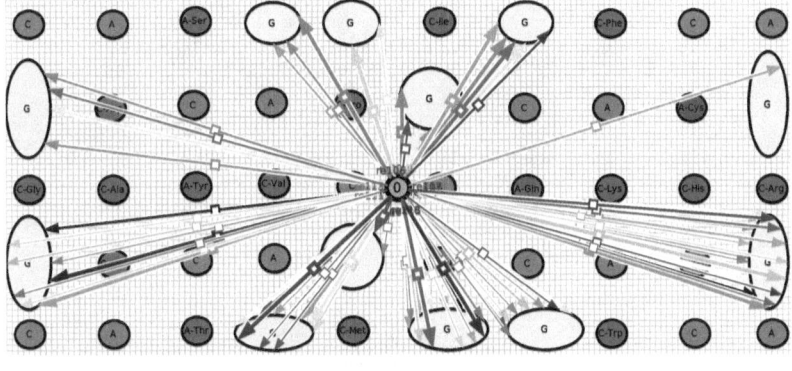

Orthogonal triangular projection

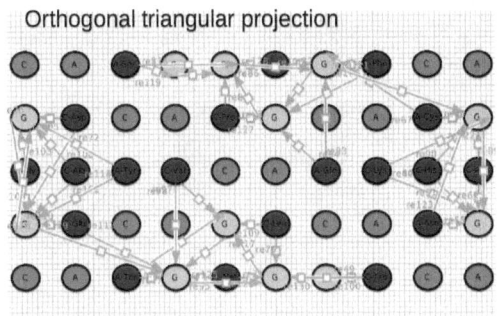

---

[1]Please note the 16 arrows pointed towards right-lower quadrant

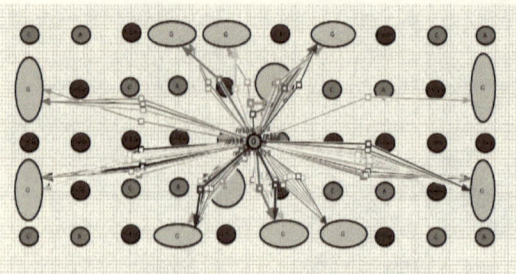

Figure 1: Hierarchic model of projections of original 5x10 matrix

2 Projections...projected

"Spider"[hierarchic]model

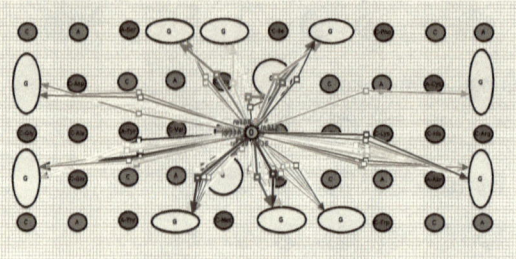

Figure 2: Orthogonal projection of original 5x10 matrix

"Inversed" distribution model

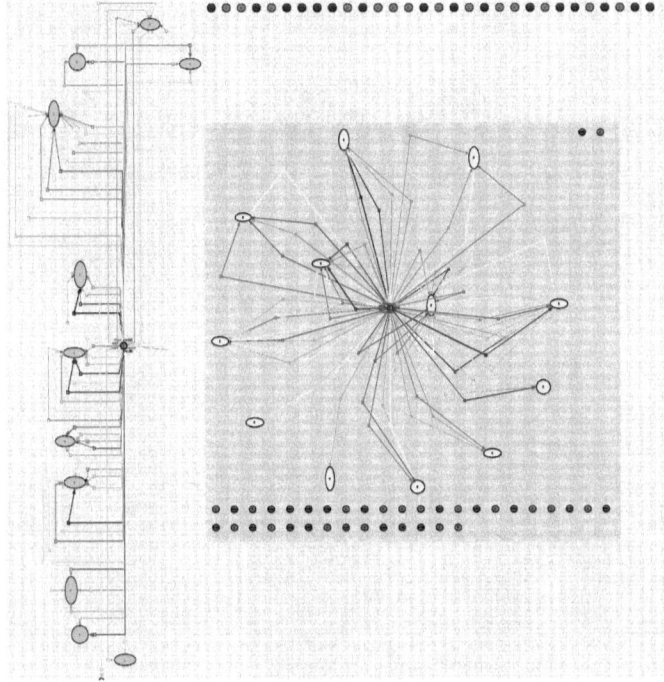

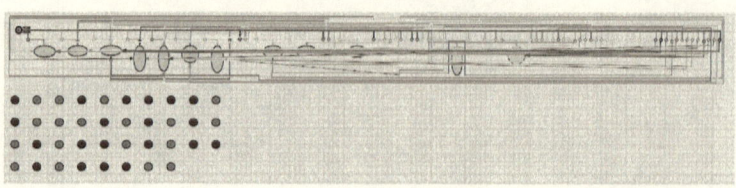

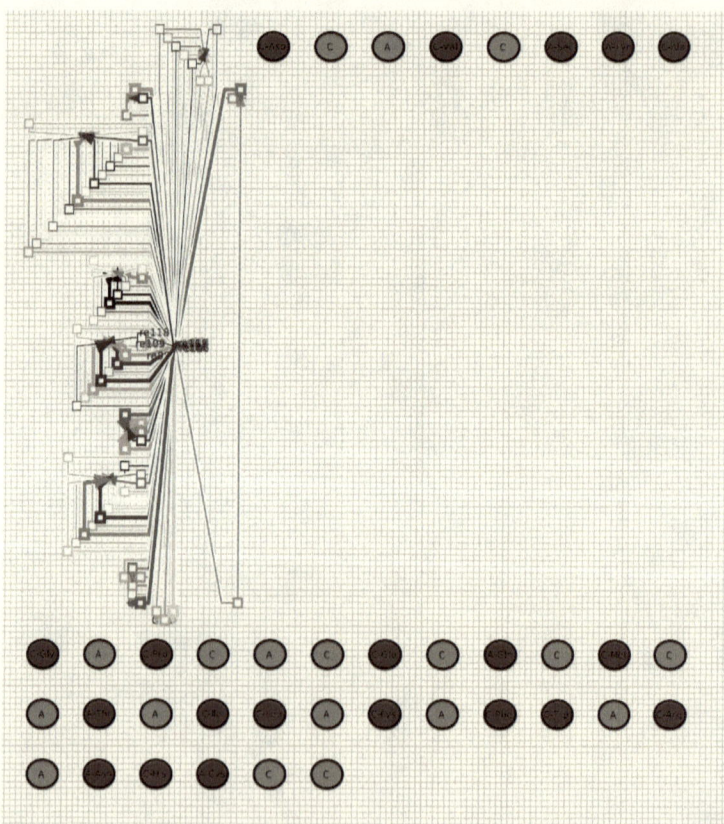

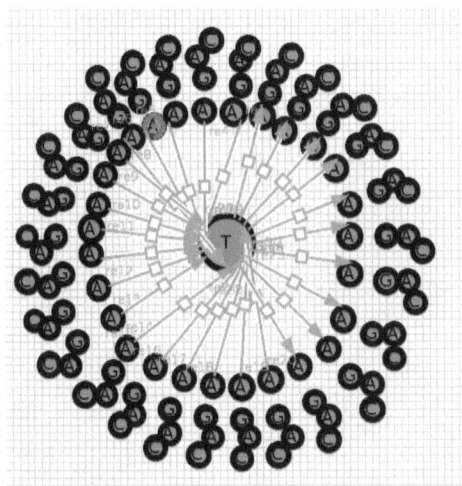

Part III
"Peusjak" model

3   Transgressive vertiginous concept

ϕ-orbitaloid aspects disperse unequilocally.

3.1   CAGAT-construct

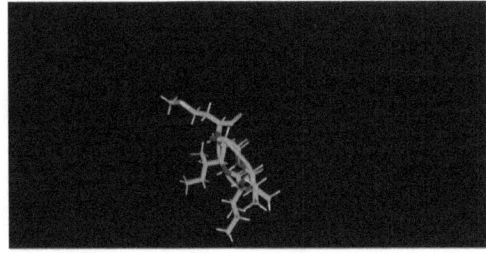

4 g-orbitaloid tripled rotatorious tensors

4.1 3δ context creates recursive movements

$$^\gamma \triangleright^{\gamma\gamma}_\gamma \triangleleft^{\gamma\gamma}_\gamma \triangle^{\gamma\gamma}_\gamma \, \triangleright^{\gamma\gamma}_\gamma \triangledown^{\gamma\gamma}_\gamma \triangleleft \,^{\gamma\gamma}_\gamma \triangleright^{\gamma\gamma}_\gamma \triangledown^{\gamma\gamma}_\gamma \triangleleft \,^\gamma_\gamma$$

Curved circular vectors: $|\frac{1}{2}c|^2 \to |\frac{3}{2}C|^2$

$$^\gamma \triangleright^{\gamma\gamma}_\gamma \triangleleft^{\gamma\gamma}_\gamma \triangle^{\gamma\gamma}_\gamma \, \triangleright^{\gamma\gamma}_\gamma \triangledown^{\gamma\gamma}_\gamma \triangleleft \,^{\gamma\gamma}_\gamma \triangleright^{\gamma\gamma}_\gamma \triangledown^{\gamma\gamma}_\gamma \triangleleft \,^\gamma_\gamma$$

Time intervals' ratios are constant, movements are equal and synchronous. Inherent fixation follows every single movement, while intervals of freedom designate significant orientation relative to the other two vectors, while $\Delta\nu = const$.

Rotations are dioagonal to the field. Overlaps are periodic, whilst movement is unidirectional.

4.1.1 New chronogenic concepts [NCC]

Resultant vector provides axonemic guidance for enrichment to the anterograde extremes.

$$^{CA}C^{G^C}AT \tag{1}$$

Centrifugal acceleration is bipedal loop. The velocity leverage is secondary. Positions determine structure similar to "paper windmill", traveling alongside the stream. Evasive circulatory branching appears evident. Important notion: Circulation is given birth by the appearance of gero($\Gamma\rho$)active differentiation stadium.

5 Static h-orbitaloid knit

5.1 Positioning random aparatus

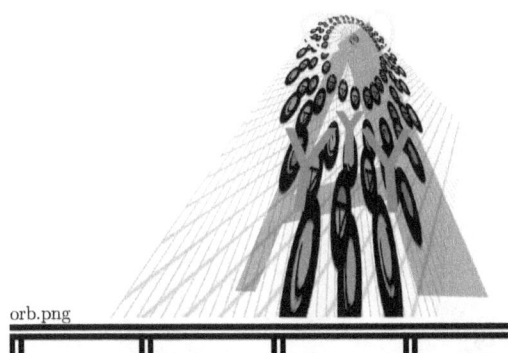

orb.png

^{1}C	^{2}C	^{3}C	^{4}C	A	^{5}C
^{6}C	A	$^{7}C^{1}$	^{8}C	A	^{9}C
^{10}C	^{11}C	A	A	$^{12}C^{2}$	^{13}C
$^{1}C^{3}$	^{2}C	A	$^{3}C^{4}$	A	$^{4}C^{5}$
A	A	^{5}C	A	A	A
$^{6}C^{6}$	$^{7}C^{7}$	$^{8}C^{8}$	$^{9}C^{9}$	A	$^{10}C^{10}$

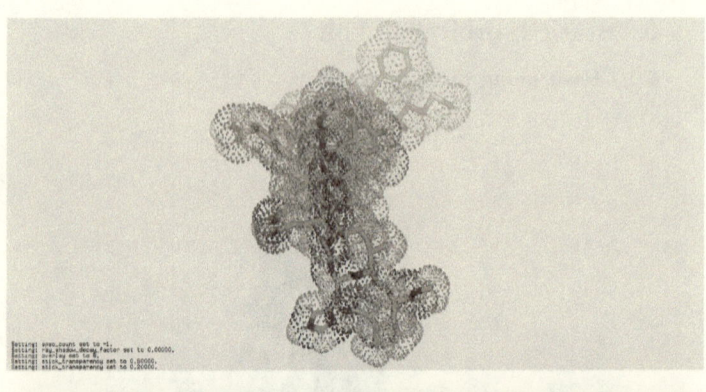

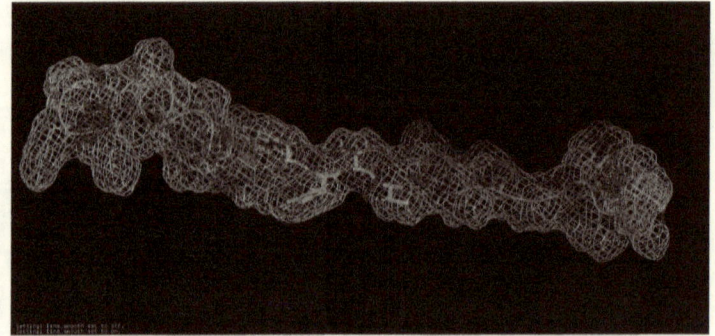

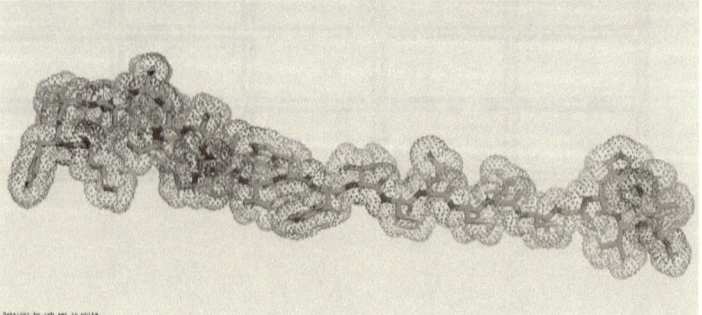

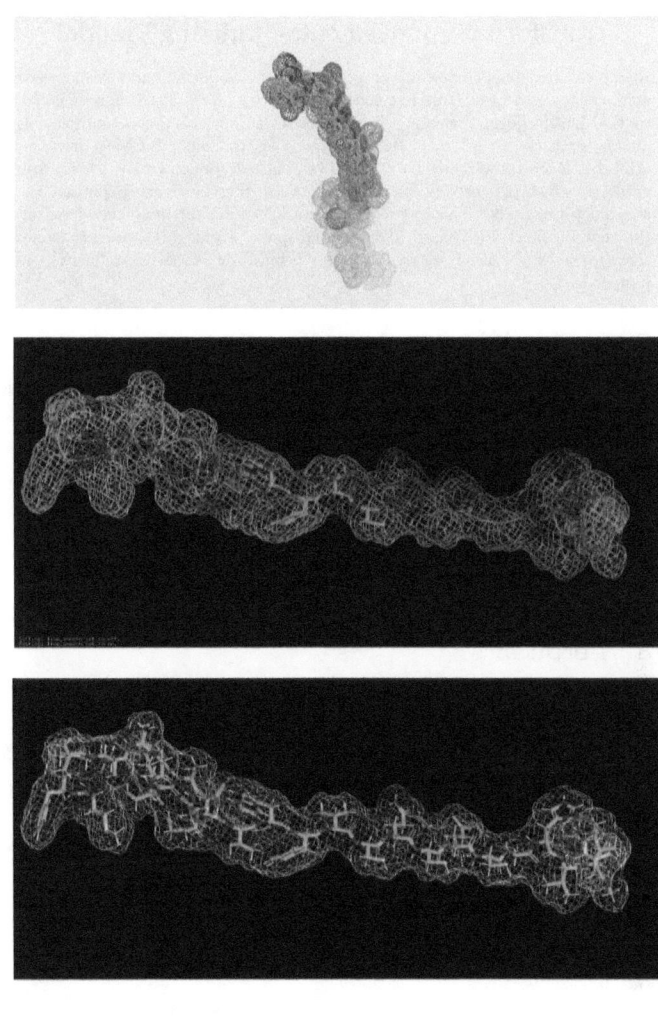

6 h-orbitaloid biphasic serial and shifting undulatory transforming movements-"knitting" model

Relatively changing planatorious proportions are being underpinned by successive axial violations, which return in convergent ankyring order. Traction is diagonal so the depth equals 2 chains. Torsion is retracting space commodities for model creation in i-orbital. But, what creates torsion? h-orbital from g-orbital or h-orbital solely, or both? Once established, torsion creates temporary gradients which pin-through the knitted network, thereby creating contexts for adjacency protocols. The active phase reutilizes i-orbital, only the shape of the hook("chingelakija") marks κ-orbitaloid space. l-orbital reappears through thread, μ orbital consist "sleeves", while ν orbital are "socks", ω-rollneck of a jumper, etc.

7 j-orbitaloic spheric perspectives

7.0.1 Concept of paralleled moving perspectives is inferring heterogenic changing focus of potentiation.

Transparency is gradual..."it is an order" Disposal of signals follows bylayering pattern. Bilateral convexed&opposing signals to field co-emerge transgressively eleviating the sensitivity to resultant appearances. Paralleled moving axes distort relative power fields.

$$\frac{t\!\supset\!\subset\!\dashv}{(\frac{1}{10} * \leftarrow U \cap \hookleftarrow)} \tag{2}$$

8 l-orbitalics

8.1 Reversed coupling of 2 perforating soap bubbles

Symetry is being marginal while movements are diagonal. Dispersion is mutually synchronous. Common sections regress as far as directed motions are paralleled. Surface area is a function of distorsive pathways. Surface tensions create limitations which refract→reflect→refract or reflect→refract→reflect signals while angles of reflection/refraction besides being variables, follow certain distribution which is derived through the specific weight of the media.

8.2 Focus represents inversion of surface tension

Gradient circular area must not be replaced with volume-portions of diffusion.Back-focused linear projections of common sections make the linear part of l-orbitalics.
Gas diffusion is inflating, not deflatory
Π angled orientation solves the basal membranous exchange of gasses.

9 Surface&volume friction basics

Self propagating μ-orbital U shaped velocity creates the existence of μ-orbital. Following laws meant for kinetic energy, it's machinery is bipahse-limited, one time inductively, followed by reversed deductive trajectory. Flux comes first, whereas undulation goes on. It creates context for circulation.

$$n T n \qquad\qquad (3)$$

Part IV
Continuous voltage&spiky amplitudes(Biphased n-orbital)

Potentiated capacity raises approximation over extreme changes in flow. Therefore spikes are adopted in the context of comparative analysis over field continuum. It is inherent, but relative. In biomedicine cuboid and cylindrical epitel, conforming the frontline in soft tissues, its relative existence to squamous epitel in particular, illustrates vividly the very transition from m to n orbital, weather the reverse pathway becomes regenerative. Spikes are reactive and obturatory. Their appearance is arbitrary, while their confluence is layered.

$$u^{\cap}\cap^{u} \qquad\qquad (4)$$

Note the letter Z(cyrilic:)
Cross-projections of n-orbital are approximations.

Equalization is force-oriented[Liberal O-orbital]

$$uT\imath T{\leftrightarrow}U \qquad\qquad (5)$$

Direction is important whenever energy transfer is considered[p–orbital] The possibility for existence of π–orbital being non-recursive, becomes new hallmark for circuit of cell currents[2]
Molecules are as far as one can get in terms of π-crossings. In the context of π–crossings, molecules are represented in cell currents themselves[3].

---

[2]the formulation "cell circuit of currents" is posed from the observer's perspective.
[3]See: New Essentials in Genetics:Methionine is π, Dr. Kliment Sandjakoski, 2017, LAP

Part V

Doubled persisting&recircling P orbital

$$CAU \qquad\qquad (6)$$
$$AUG \qquad\qquad (7)$$

Self distorsion bundles on the knitting model. However the k orbital is not flexible, therefore motion promotes II orbital.

The P orbitalic outburst, by definition is the right&left arm of Eindhoven's triangle.

In the case of asymetry DC becomes gradual.

2

THE GENETIC CODE

Genetic code is meant to code the genetic information into series of three letter blocks that both in a continuous and discrete manner determine cell fate. It is a four letter code which serves to install a genetic program including numerous processes we know very little about. However, the variations in the four letters are not the sole information that are, simply put, only one side of the medal. The code is not only letters, but numbers as well.

2.1 Introduction

It's a code. It is numbers as well. Profound understanding of the four letter concept in the creation of the genetic code lies in transforming the language in such way that it uncovers fundamental scientific facts regardless of their present potentials and/or structure. Nevertheless, since particular scientific field have become exhausted, there seem to be no more accidents. No mistakes, neither experiments could

INTRODUCTION   21

overturn the outcome of the learning process that has become robust and in well accordance with matching, previously established, scientific facts. However, the context is a changing matter, that applies to knowledge as such, in a discrete and principal manner. Those mechanisms differ and serve to lead scientists over the centuries to discover new frameworks, postulates, principles, laws and formulas as to articulate them comprehensively for more general conditions.

This book represents new concepts, formulas, principles, so it creates platform for further science efforts concerning the genetic code. The work is originally created by the author himself, it is inconclusive unless the reader engages into a systematic analysis, brainstorming the infinite number of rules and exceptions that may arise due to the prospected model or a situation.

Sequencing by hand

The genetic code is programmable. It represents a language that is made of basic symbols(e.g. A, C, T, G, U) and certain rules apply in order to produce meaningful coding sequence. This may infer the existence of a transforming power of the signal to the point where its actual activity produces observable outcome. Therefore the combination of transcription and translation process that gives rise to the "central dogma in biology" is to discriminate in time-mode(plot where one of the axes is timescale) the realization of the genetic information.

Herein the radically new idea would be to understand transcription not primarily as quantitative and transmitting process of the original DNA-coding signal(DNA sequence) to "ribosome factories" in the cytoplasm, but rather as a selection process that extracts and sets the limits to the coding sequence apart from the surrounding nucleotides in the predesignated order. The radical change in the emphasized understanding is the fact that both, transcription and translation, represent a single transformation process of deposited genetic information in the DNA to generate specific change in the cell.

Fourier transform is well-studied mathematical concept that could be used to explain existing analogies in this context. There is also a modification of Fourier transform that imposes limits for noncoherent[4] waves that could be turned into coherent for

---

[4]not coherent

INTRODUCTION   23

limited time intervals so they could be applied fourier
transform. There is some risk of distortion physics
if this limits are not well set. The analogy is that
transcription is analogous to the limitation step in
the, lets say not coherent wave called DNA and
the actual sampling of the extracted genetic information to the limit where all detectable frequencies
are reproduced(splicing etc.). Translation is a process
where once dissolved waves that had been part of the
selected complex wave called DNA-sequence are now
utilized again to reconstruct a signal that corresponds
to the original signal but is at the same time much
less complex and bulky.

Furthermore, major consequences could be inferred
from this new understanding of transcription in
regard to the expression levels, as researchers determine, relative to certain genes. *Expression levels of a
gene are a measure of the complexity of genetic information carried out by the transcribed gene or DNA
sequence.*

Finally, genes are transcribed simultaneously or in
a timely fashion where they follow certain principles and rules. In that process they create different

amount of transcripts. Therefore it is absolutely possible to reconstruct genetic information with respect to its complexity and (non)coherence.

Two methods are well known and established in the science world today. One of them is highly performed in the latest advances concerning the human genome project. However this part of the book is meant to show as there may exist alternative methods or predicaments in the approach to the actual order of nucleotides in the structure of the important nucleic acids(DNA, i/mRNA, tRNA, rRNA, mitDNA, etc.)

2.2 Methionine is "π"

One of the very basic findings widely entertained in biomedicine or any related field is the start point of every translational process. Methionine has the reserved first position on every product of translation[5]. On the scheme, created and designed below, my intention was to shed light on quite, entirely new and very important conclusions.

---

[5]Some authors claim there are minor alterations to this principle

METHIONINE IS "π"   25

Scheme Description

The idea is to classify all of the amino acids that do not require specific nucleotide for its coding into same group. Therefore four groups may exist:

   -A(no A), -G(no G), -C(no C), -T(no T)

Scheme Analysis

Table 1: Scheme Presentation

-A	-G	-T	-C
Phe	Phe	Pro	Phe
Leu	Leu	His	Leu
Ser	Ser	Gln	Tyr
Cys	Tyr	Arg	Cys
Trp	Leu	Thr	Trp
Leu	Pro	Asn	Ile
Pro	His	Lys	Met
Arg	Gln	Ser	Asn
Val	Ile	Ala	Lys
Ala	Thr	Asp	Ser
Gly	Asn	Glu	Arg
....	Lys	Gly	Val
....	....	....	Asp
....	....	....	Glu
....	....	....	Gly

The foremost finding on the presented scheme is the position of methionine(commonly referred to as the starting point of every translational process). Its position on the scheme renders this amino acid special in terms of the code-structure, that solely encodes this particular amino acid. It must contain all the three nucleotides: A, G and T and at the

same time the code must not contain the nucleotide
C. Only those nucleotide triplets which meet both
criteria(mentioned above) has the ability to encode
methionine. *Methionine is the single unique amino
acid with such encoding property.*

Another important conclusion from the analysis is
rather obvious finding that any amino acid can be
encoded without the need for cytosine to take part
in the nucleotide triplet. This holds true only with
cytosine.

2.3 Code is just code, then code is patterns, then code is just code again

Proceeding further in the process of reconstructing
the genetic code into an elementary model, it is legit-
imate to question and to reestablish the conditions
under which 1, 2 or 3 identical nucleotides are being
organized in triplets therefore defining the coding
potential for specific amino acids.

Defining Least Common Denominator (LCD)

Genetic code in literature is frequently referred to
as degenerated. Thus in the context of continuous nucleotide sequence, presumably there should

CODE IS JUST CODE, THEN CODE IS PATTERNS, THEN CODE IS JUST CODE AGAIN

exist a minimum data required about the designated sequence and a set of principles, that would be sufficient to reconstruct the exact order of nucleotides or exhaust its coding potential in an alternative pathway. Such algorithm will enable to look through possible key principles in putting together nucleotide triplets or nucleotides in a designated DNA sequence.

In Searching for Least Common Multiple (LCM)

Is there a threshold value pointing towards a switch from quality to quantity in the coding process. Does such switch exist? Is quantity coded in a nucleotide sequence. This work is intended to set up an environment for direct research on original ideas considering theoretical analysis of the genetic code and its implications.

Decoding algorithm of the genetic code (Kliment Sandjakoski)

Decoding formulas

Kliment Sandjakoski

There are three steps in this process. In the first step we will try to reformulate the order set bu the coding rules, so we could analyse them from different perspective. Thereafter, in the second step, an extensive analysis will take place so that we can finally encircle our findings with the well known facts from classical molecular biology principles.

Middle Base / 5' Base	U	C	A	·G	Middle Base / 3' Base
·U	Phe	Ser	Tyr	Cys	U
	Phe	Ser	Tyr	Cys	C
	Leu	Ser	Stop	*Sec Stop }	·A
	Leu	Ser	Stop	Trp	G
C	Leu	Pro	His	Arg	U
	Leu	Pro	His	Arg	C
	Leu	Pro	Gln	Arg	A
	Leu	Pro	Gln	Arg	G
A	Ile	Thr	Asn	Ser	U
	Ile	Thr	Asn	Ser	C
	Ile	Thr	Lys	Arg	A
	▲ Met Initiator }	Thr	Lys	Arg	G
G	Val	Ala	Asp	Gly	U
	Val	Ala	Asp	Gly	C
	Val	Ala	Glu	Gly	A
	Val	Ala	Glu	Gly	G

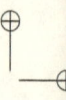

1.1. Pandora's box

Criteria for most diversely coded aminoacids:

1. Phe=-A;-G;
2. Glu=-C;-T;
3. Trp=-C;-A
4. Lys=-C;-T

1. Asn=-G;
2. Tyr=-G;
3. Cys=-A;
4. Ile=-G;
5. His=-G;
6. Gln=-T;

1.2. Degrees of freedom for aminoacids

1.2.1. Primary aminoacids

Pro CC*
Thr AC*
Ser= UC* AG*
Leu= CU* UU*
Val= GU*
Ala= GC*
Gly= GG*
Arg= CG* AG*

1.2.2. Secondary aminoacids

Phe= UU*'
Ile= AU*'
Lys= AA*'
Glu= GA*'
His= CA*'

1.2.3. Tertiary aminoacids

Asn= AA*"
Cys= UG*"
Tyr= UA*"
Asp= GA*"
Gln= CA*"

1.2.4. Fixed aminoacids
Trp= UGG
Met= AUG

1.3. Degrees of freedom for codons
- CC
- CG
- AC
- GU
- GC
- GG
- CU
- UC
- UU′′
- AU′′
- AA′″
- GA″″
- UA″
- UG″″
- CA″″
- AG″″

1.4. Summing up degrees of freedom
2+2+3+4+2+4+4+4=25

$\sqrt[3]{25} = 5$

1.5. Conclusion
There are 5 nocleotides and there are total 5 bonds between nocleotides(double(A->T) +triple(C->G))

2.4 Does degeneracy of the genetic code exist?

"Degeneracy of codons is defined as redundancy of the genetic code, exhibited as the multiplicity of three-base pair codon combinations that specify an amino acid."

The very essence of this book entertains the idea that redundancy in codons is giving out key secrets to decipher their true role in the human genome. Therefore the concept of unraveling the role of each nucleotide at a unique position in DNA contradicts the idea of degeneracy. Another idea was that repetitive behavior is impossible in the context of two different codons at various locations in the genome which code the same amino acid. Simply put, knowing the terms in a language does not impose we understand the true meaning of its sentences.

2.5 Coding the language or decoding the code

The genetic code is historically postulated as one of the numerous contemplating efforts to explain the molecular organization of DNA. It turned out that nucleic acids are far more complex than researchers expected, so biomedical science required the overcome of another important milestone, namely the Nobel prize winning concept of sequencing(early 70s). Thirty years later researchers were able to completely sequence the human genome. The method was established, the concept was shown correct. What now? Have we reached a dead end? We read through it as we try to understand the signs from predecessors, millions of years ago.

What if DNA is language? Fluid, unstable, interchangeable, contemporary language. Then we would probably be invited to speak and write that language, understand its formulations and discover its existing interpretations, finally communicate on that language. This is one of the very few fundamental standings in this book concept wise so significant volume of pages is fulfilled with principles and grammar of the prospected language[6].

2.6 Molecular characterization of nucleotides

Atomic structure of nucleotide-molecules reflect on it's functional role in the organization of DNA. The root to their behavior in a context of a genetic sequence sprouts from atoms, also their localization and bonds in the nucleotide. The specific part in the nucleotide is the nitrogen base, which may

---

[6]For the purpose of this book we may call it DNA-language

MOLECULAR CHARACTERIZATION OF NUCLEOTIDES 33

purine(A&G)or pyrimidine(C&T). Purine bases consist of five nitrogen atoms four of which are encircled in aromatic rings and one that proxies to different position in the ring. Guanine-purine ring on the 6th position binds oxygen with double covalent bond. Notably nitrogen atoms have the 1st, 3rd, 7th and 9th position in the guanine aromatic rings. Arising from the 2nd position(C-atom) is single covalent bond to proxied nitrogen atom in NH2 group. In the case of adenine, the single covalent bond to NH2 group is positioned at 6th position, having nitrogen atom on the 1st and carbon atom on the 5th position in the ring. Nitrogen atoms have the same distribution in the adenine purine-ring as in the guanine-purine ring(1st, 3rd, 7th and 9th position). Pyrimidine nitrogen bases also have certain regularity in their structure that predefined their behavior in particular DNA-sequence. Cytosine is organized as single nitrogenous aromatic ring with nitrogen atoms on 1st and 3rd position and double covalent bond pointing from 2nd position towards oxygen atom. Another nitrogen atom as part of -NH2 group lies outside the ring bonded to its 4th

34   THE GENETIC CODE

position by single covalent bond. Whereas cytosine
has three nitrogen atoms in its structure, thymine has
only two, on the 1st and 3rd position in the ring. Carbon at 2nd and 4th position covalently bind oxygen
with a single bond. There is methyl group covalently
bound the C-atom at the 5th position in the ring.

G
Guanine

A
Adenine

C
Cytosine

T
Thymine

It is no coincidence that only these characteristics in
the molecular structure are mentioned in the text.
They specifically relate to molecular characterization of the micro context in certain sequence or
region in DNA. In this regard, it is worth mention-
ing another postulate, namely that dominant and
recessive behavior of genes is coded in the molecular structure of nucleotides and the order of their
sequence. Moreover the sensitivity and specificity to
reproductive signals, as a constitutive attribute per
se, is also a matter of the nucleotide-quality of the
sequence.

The explanation of how chemical structure of
nucleotides predefines the outcome in terms of genetic
behavior, I chose to present elsewhere.

2.7 Nucleotides phase shift

Fourier transform of the genetic signals

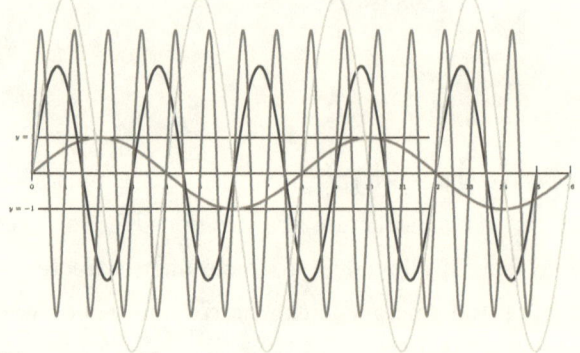

Earlier in the book a radically new idea was intro-

duced in what was widely known as transcription
and translation and their possible unification in a
transformational process based on Fourier transform.
Herein it is important to focus on culprits which give
rise to such an idea.

Assuming the concept of wave-particle duality,

nucleotides may be regarded as waves that have constant specific properties. As such processes as interference, diffraction and dispersion are inherent to their behavior as waves. As such, nucleotides interact with adjacent nucleotides depending on the qualitative properties of the DNA-chain. Those interactions occur in specific phase points which are determined by the two adjacent nucleotides in the DNA-chain that interact, as well as the surrounding DNA-sequence. Nevertheless these interactions tend to be fixed in phase, due to the conserved structure of the DNA chain itself. However, phase shifts exist as part of the whole set of changes that permanently occur and range from physiological to pathological for the cell. Those shifts obviously tend to have limits in order to preserve the primary structure of DNA and that is its nucleotide sequence.

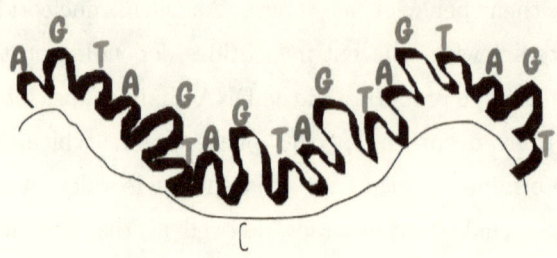

2.8 RNA–vocabulary

RNA-subdivision into multiple functional entities regarded as separate RNA-s is fundamental concept theoretically postulated from observations prior to their isolation in experimental conditions. Another fact speaks in favor of the existence of robust formula, certain regularity in the RNA-structure and that is evolutionary standings pointing out RNA as predecessor to DNA and proteins. Therefore RNA should not be only regarded as simple transcript or amplifier in between DNA and proteins but rather as an evolving feature of nature that was subdivided in different types each of them existing in its own system and executing their separate predefined actions.

Hereby it is entertained how continuous analysis of RNAs cell-system leads us to the fundamental sprouts of RNA-vocabulary.

2.9 RNA-grammar

Certain principles must apply to the four-letter code to put it in a sensible order. Those principles derive from essential functional roles these nucleic acids play in the cell[a]. RNA represents the in-between stage in the central dogma of biology:

$DNA- >< -RNA- > protein.$

In an age of mastering RNA-seq technologies, detection characterization and utilization of new types of RNA as well as ex vivo synthesis of such for the purpose of experimental techniques occurs every few years. The image below shows us the four basic nucleotides that conform the structure of essentially all types of RNA.

---

[a]considered by its basic definition as elementary functional and structural unit of living organisms

Solution ...UAGCAUAAAAAAA...

---

m/iRNA–syntax

Solution $UUU\_UUU\_UUU\_U$

---

m/iRNA Syntax

1 ACAUCACUCCCUG   61
2 CCAUAACUGCCUC 32   62
3 CCCUAAAUCGCUC 33   63
4 CACUACAUCCGUC 34   64
5 CAAUACCUCCCUG 35   65
6 AACUCCAUCCCUG 36   66
7 CCCUAAAUCCCUG 37   ACAUGACUCCCUC 67
8 AAAUCCCUGCCUC 38   ACAUCGCUGCCUA 68
9 CACUACAUGCCUC 39   ACAUCAGUCGCUC 69
10 ACCUCAAUCCGUC 40   GCAUCACUCCCUA 70
11 AACUCCAUCGCUA 41   AGAUCACUCCCUC 71
12 CAAUACCUCGCUC 42   ACGUCACUCCCUA 72
13 ACAUCACUCGCUC 43   CCAUCACUGCCUA 73   CAAUGCCUCCCUA
14 ACCUCAAUCCGUC 44   CCAUACCUGCCUA 74   CAAUAGCUCCCUC
15 CCAUAACUCCGUC 45   CCAUAACUGCCUC 75   CAAUACGUCCCUC
16 AAAUCCCUCCGUC 46   CCAUAACUGCCUC 76   GAAUACCUCCCUC
17 AAAUCCCUGGGUA 47   CCAUAACUGCCUC 77   CGAUACCUCCCUA
18 CCCUAAAUGGGUA 48   CCCUAACUGCCUA 78   CAGUACCUCCCUA
19 GGGUCCCUAAAUA 49   CCCUCAAUCGCUA 79   AACUGCAUCCCUC
20 AAAUGGGUCCCUA 50   CCCUACAUCGCUA 80   AACUCGAUCCCUC
21 AAAUCCCUGGGUC 51   CCCUAACUCGCUA 81   AACUCCGUCCCUA
22 CCCUAAAUGGGUC 52   82   GACUCCAUCCCUA
23 GGGUCCCUAAAUC 53   83   AGCUCCAUCCCUA
24 AAAUGGGUCCCUC 54   84   AAGUCCAUCCCUC
25 AAAUCCCUGGGUG 55   CACUCCAUCCGUA 85   CCCUGAAUCCCUA
26 CCCUAAAUGGGUG 56   CACUACAUCCGUC 86   CCCUAGAUCCCUA
27 GGGUCCCUGGGUG 57   CACUACCUCCGUA 87   CCCUAAGUCCCUA
28 AAAUGGGUGGGUG 58   CACUACAUCCGUC 88   GCCUAAAUCCCUC
29 AAAUAAAUAAAUG 59   CCCUACAUCCGUA 89   CGCUAAAUCCCUC
30 CCCUCCCUCCCUG 60   CACUACAUCCGUC 90   CCGUAAAUCCCUC

91		123		155	
92		124		156	
93		125		157	
94		126		158	
95		127		159	
96		128		160	
97		129		161	
98		130		162	
99		131		163	
100		132		164	
101		133		165	
102		134		166	
103	AAAUCCCUGCCUC	135	AACUCCAUCGCUC	167	ACCUCAAUCCGUC
104	AAAUCCCUGCCUC	136	AACUCCAUCGCUC	168	ACCUCCAUCCGUA
105	AAAUCCCUGCCUC	137	AACUCCCUCGCUA	169	ACCUCACUCCGUA
106	CAAUCCCUGCCUA	138	CACUCCAUCGCUA	170	CCCUCAAUCCGUA
107	ACAUCCCUGCCUA	139	ACCUCCAUCGCUA	171	ACCUCAAUCCGUC
108	AACUCCCUGCCUA	140	AACUCCAUCGCUC	172	ACCUCAAUCCGUC
109	CACUCCAUGCCUA	141	CAAUCCCUCGCUA	173	CCAUACCUCCGUA
110	CACUACAUGCCUC	142	CAAUACCUCGCUC	174	CCAUAACUCCGUA
111	CACUACCUGCCUA	143	CAAUACCUCGCUC	175	CCAUAACUCCGUC
112	CACUACAUGCCUC	144	CAAUACCUCGCUC	176	CCAUAACUCCGUC
113	CCCUACAUGCCUA	145	CCAUACCUCGCUA	177	CCAUAACUCCGUC
114	CACUACAUGCCUC	146	CACUACCUCGCUA	178	CCAUCCCUCCGUA
115	ACCUCAAUGCCUC	147	ACAUCACUCGCUC	179	AAAUCCCUCCGUC
116	ACCUCCAUGCCUA	148	ACAUCCCUCGCUA	180	AAAUCCCUCCGUC
117	ACCUCACUGCCUA	149	ACAUCACUCGCUC	181	AAAUCCCUCCGUC
118	CCCUCAAUGCCUA	150	CCAUCACUCGCUA	182	CAAUCCCUCCGUA
119	ACCUCAAUGCCUC	151	ACAUCACUCGCUC	183	ACAUCCCUCCGUA
120	ACCUCAAUGCCUC	152	ACCUCACUCGCUA	184	AACUCCCUCCGUA
121		153		185	
122		154			

Table 1: Translational Medical Research

Manufactured by: Hikmet Savit(Istanbul, MD)

TMR program

DNA-GRAMMAR

tRNA–syntax

Solution CGG-U-GGG-U-UAU-U-UUU

---

rRNA–syntax

	C	G	C	C	C	G	G	C	C	U	G	C	
C													A
G		C											C
C			C										G
C													A
C										A			A
G			G		C								C
G						C							C
C													A
C													A
C								G					A
G											G		C
C													A
	U	G	U	U	U	G	G	U	U	C	G	U	

2.10 DNA-grammar

DNA grammar is of different quality, based on the very concepts of its existence. It is continuous series of nucleotides that demonstrate enormous variations, divergence and numerousness.

44   THE GENETIC CODE

mtDNA–syntax

®Prospected structure of mitochondrial DNA

```
TAC GTT TAC GTA TCG TAC TGT ACG        TTA CGT ATA CGT CTC GTA GTG TAC
ATG CAA ATG CAT AGC ATG ACA TGC        AAT GCA TAT GCA GAG CAT CAC ATG

TTT ACG TAT ACG ACT CGT CGT GTA        GTT TAC GTA TAC TAC TCG ACG TGT
AAA TGC ATA TGC TGA GCA GCA CAT        CAA ATG CAT ATG ATG AGC TGC ACA

CGT TTA CGT ATA GTA CTC TAC GTG        ACG TTT ACG TAT CGT ACT GTA CGT
GCA AAT GCA TAT CAT GAG ATG CAC        TGC AAA TGC ATA GCA TGA CAT GCA
```

DNA–syntax

Comprehensive DNA sequence.
ACGGTCAGCTTCGCAACGTCAGCTTCGCAGCACTGCACGTCCC
TGCAGTCGAAGCGTTGCAGTCGAAGCGTCGTGACGTGCAGGG

CCCCCCCCCCCCCCCAAAAAAATTTTTTT
GGGGGGGGGGGGGGGTTTTTTTTAAAAAAA

CGCGCGCGCGCGCGAAAAAAAAAAAAAAA   ACACACA
GCGCGCGCGCGCGCTTTTTTTTTTTTTTT   TGTGTGT

CACACAC   TGGCTGCTACCCG
GTGTGTG   ACCCAAGGTGGCC

2.11 Rationale for codifying genetic language

Sequencing formula

$$atatatata'atccatg''c'G,,,$$

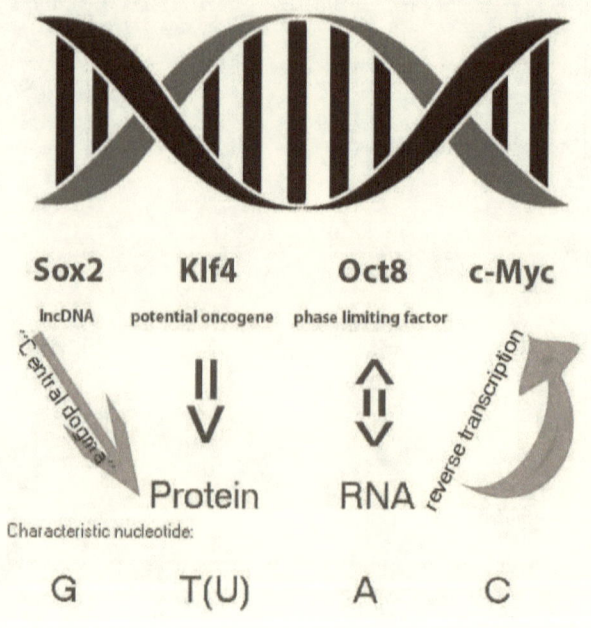

The triangle paradigm

Reprogramming target genes

Referring to the recent discovery of four crucial genes in dedifferentiation to hipSCs

$(Sox2)_3 - > Gly$ $(Klf4)_3 - > Phe$ $(Oct3)_3 - > Lys$ $(c\text{-}Myc)_3 - > Pro$

$(Sox2)_2Klf4 - > Gly$	$Klf4(Sox2)_2 - > Trp$	$Sox2Klf4Sox2 - > Val$
$(Sox2)_2Oct8 - > Gly$	$Oct8(Sox2)_2 - > Arg$	$Sox2Oct8Sox2 - > Glu$
$(Sox2)_2c\text{-}Myc - > Gly$	$Sox2(Klf4)_2 - > Arg$	$Sox2c\text{-}MycSox2 - > Ala$
$(Klf4)_2Sox2 - > Leu$	$Oct8(Klf4)_2 - > Val$	$Klf4Sox2Klf4 - > Cys$
$(Klf4)_2Oct8 - > Leu$	$c\text{-}Myc(Klf4)_2 - > Ile$	$Klf4Oct8Klf4 - > Tyr$
$(Klf4)_2c\text{-}Myc - > Phe$	$Sox2(Oct8)_2 - > Leu$	$Klf4c\text{-}MycKlf4 - > Ser$
$(Oct8)_2Sox2 - > Lys$	$Klf4(Oct8)_2 - > Glu$	$Oct8Sox2Oct8 - > Arg$
$(Oct8)_2Klf4 - > Asn$	$Klf4(Oct8)_2 - > I$	$Oct8Klf4Oct8 - > Ile$
$(Oct8)_2c\text{-}Myc - > Asn$	$c\text{-}Myc(Oct8)_2 - > Gln$	$Oct8c\text{-}MycOct8 - >Thr$
$(c\text{-}Myc)_2Sox2 - > Pro$	$Sox2(c\text{-}Myc)_2 - > Ala$	$c\text{-}MycSox2c\text{-}Myc - >Arg$
$(c\text{-}Myc)_2Klf4 - > Pro$	$Klf4(c\text{-}Myc)_2 - > Ser$	$c\text{-}MycKlf4c\text{-}Myc - >Leu$
$(c\text{-}Myc)_2Oct8 - > Pro$	$Oct8(c\text{-}Myc)_2 - > Thr$	$c\text{-}MycOct8c\text{-}Myc - >His$

$Sox2Klf4Sox2 - >Val$	$Klf4Oct8c\text{-}Myc - > Tyr$	$Sox2Oct8c\text{-}Myc - >Asp$
$Sox2Oct8Klf4 - >Asp$	$Sox2c\text{-}MycOct8 - > Ala$	$Sox2c\text{-}MycOct8 - > Ala$
$Oct8Klf4Sox2 - > Met$	$c\text{-}MycOct8Klf4 - > His$	$Oct8Sox2c\text{-}Myc - > Ser$
$Oct8Sox2Klf4 - > Ser$	$Oct8c\text{-}MycKlf4 - > Thr$	$Oct8c\text{-}MycSox2 - > Thr$
$Klf4Sox2Oct8 - > I$	$Oct8Klf4c\text{-}Myc - > Ile$	$c\text{-}MycOct8Sox2 - > Gln$
$Klf4Oct8Sox2 - > I$	$Sox2Oct8Sox2 - > Glu$	$c\text{-}MycSox2Oct8 - > Arg$

$c\text{-}MycKlf4Sox2 - >Leu$	$Klf4c\text{-}MycSox2 - > Ser$
$c\text{-}MycSox2Klf4 - > Arg$	$Sox2c\text{-}MycKlf4 - > Ala$
$Klf4Sox2c\text{-}Myc - > Cys$	$Sox2Klf4c\text{-}Myc - > Val$

2.12 Phase limiting vs. time limitations

Time limits are as fundamental as the concept of time. They could be set, managed, changed and challenged by the observer and his point of view. Time is one of the seven SI base units, so from physics' standpoint it is indefinable. Today we know as time differs in different parts in space, the same did different parts in physics. Namely in relativistic physics there is no absolute time as the Newtonian classical idea of time, moreover there is indivisible space-time construct. However it is clear that time limits exist naturally, the job is to find it in the context of our present scientific or other interest.

The formation of wave is grounded in the concept of phase-shifting. In idealistic physical conditions phase shift would be a constant, but in reality processes spread by different magnitudes and velocities. Dual nature of matter , both corpuscular-discrete and continuous wave-like was postulated at the beginning of the 20th century. Different phases of a single wave could therefore reach one-another or be driven apart almost all the time rather than maintain constant distance as they would in idealistic conditions. When they strive to overlap the precursor phase becomes upper limit, whereas when driven apart follower phase becomes lower limit, which creates an interval from " lowest to the utmost limit". This diapason has certain regularity represented by a factor called phase limiting factor.

The concept of Phase limits

Phase limits are waves. All physical attributes credited to waves are valid for Phase limits as well.

There must be a factor

Having properly applied time limits to even non-coherent waves we can turn them into coherent for the defined time interval. Coherent waves have certain regularity which might be formulated in a mathematical function, meaning there must be a factor(number) that factorizes the formula for the designated mathematical function.

A potential limiting factor crosses barriers of ancient biological paradigms

Limitations in the actual experimental techniques, heavily depend on day-to-day advancement of theoretical paradigms as well as the philosophical concepts. Historically, they seem consequential, as

PHASE LIMITING VS. TIME LIMITATIONS 49

opposed to the granted priority for targeted protocols. Recent development in Molecular biology is
derived from common understanding towards fron-
tier research efforts. Hereby, we present a theoretical
model for the detection of confounding data in a provisional experimental setting.

It is well appreciated, as cell progresses through its
life-cycle, that the condensation ratio in between the
discrete structures of the genome is oscillatory in
a non-coherent manner. Efficacious insight to the
reduction-driven meiosis need to encounter the logical circuit of the crossing over.

There it is likely to permute certain structure that
evades the residues of a purely confidential approach.
Nonetheless, transformation is a coupled process to a
number of predefined risk factors. Results are rather
ambivalent as cells continually regress in the differentiation process. However, molecules do determine
vibrant domains for the purpose of the processes in
cell biology. Convoluted or rectified positioning of different cell types, often being conceived as moving
gradient vector, is defining for the cell-morphology.

Nevertheless, pre–tolerated cells by the immune system, together with those MSCs that circumvented RES(reticuloendothelial system) are prone to transmigration. The conclusive explanation for it, remains elusive. Translucent efforts to characterize the cell had fostered appealing sorting measures of clinical relevance in evidence based medical practice. Electrochemical gradient's forces, being projected in between protein codominance, compromise the dissociation constant at $7.326 <= pH >= 6.736$. Ion exchange is stapling the ongoing peroxidation, as cytoskeleton structures normalize for resultant osmotic changes. Also, cellular behavior is relied on metabolism-derived coenzymes that serve to cross by the biological membranes. Altogether, versatile cell lines with potential to differentiate are supportive for the concept of a balance that is constantly changing.

Phase limiting factor

Phase in physics acquires its meaning only when correlated to a timescale. It denominates a stage in a process that is connected to a certain point on a time-axis and makes sense in the context of waves as continuous state of matter. Time has limits, so can one impose limitations to phases? This is not the essential question for the purpose of this section, but rather we discuss over the existence of a factor that imposes limitations on phases. Although phase limits might not exist, phase limiting factor exists. It is conceived as a factor which determines the quality of the change in phases that are caused by various external factors. When exactly phase changes can not be estimated, neither what quality of change is it, however phase limiting factor produces that change.

How can this factor be calculated? The factor represents a ratio of all the good things of certain phase(in numerator), divided by all the bad things of the next phase(in denominator). This factor is the same for electromagnetic waves and mechanical waves. In physics waves are well established as a product of "phase delay". The existence of phase limiting factor stands on the opposite side of "phase delay", stating that waves are a product of limiting the phase. For instance, one oversimplified example demonstrates the definition written above: when given a pendulum, it is sufficient to set narrow borders on both sides of the pendulum that are shorter than pendulum's amplitude and waves on the thread. **Solely applying borders on oscillations, produces waves.** In the context of newly produced wave, an oscillation represents a phase. Electromagnetic induction is nothing else but applying borders to electric field and an electromagnetic wave is produced.

Significance

In the context of genome regulation, the principle of phase limitations becomes essential. The four elementary constituents of DNA-the four nucleotides are combined in the primary structure of DNA. Each and every transcription represents limiting the continuous primary structure of DNA. The interpretation on these limits in possible only when relative measures are introduced.

From molecular biology perspective, in order to sufficiently explain the significance of phase limits, it is important to present new definition of the term called: "genome leverage".[7]

Redefining genome leverage: Whenever new generation of an organism is being produced through sexual reproduction, parental genomes

---

[7] It is used in different contexts in modern genetics, and thereby acquires completely different meaning

PHASE LIMITING VS. TIME LIMITATIONS   53

combine in a new "united" genome of their offspring[8]. Genome leverage occurs when certain genes inherited from one parent, combined with different set of genes of the other parent, are being accentuated or suppressed due to the inter-regulation in between genes. Whenever mendelian inheritance principles are applied, this multiplication or reduction of expressive power in the phenotype results from the unique combination of the genomes of both parents. In a purely mendelian context there are genes which will dominate in the phenotype, however the level of expression in the phenotype remains elusive. The new network of inter-regulated genes[9] will determine the power of expression in the phenotype. Therefore, gene leverage represents the difference in expressive power of the gene in the new combined genome in the offspring.

Why is introduction of a concept of genome leverage important in the context of phase limitations in the genome from the molecular biology perspective? One gene can be levered upstream(negative leverage) or downstream(positive leverage) in the (re)combined genome of the offspring. Phase limiting factor in the molecular biology perspective sets the physiological saturation point to which one gene could be levered(positively or negatively).

---

[8]In this setting it is implied that both parent organisms are viable and relatively healthy

[9]only those that are relevant to the gene of interest

From cell biological viewpoint the limitation of separate phases might be influenced by the constantly changing electro-physiological state of the cell. This phase limiting is important from the perspective of the above postulated existence of genetic language and putting in context the gene of interest.

Finally, **limitations in elongation may be regarded as transitional points from one to another wave, although these waves might be the same. Then they speak of waves' intensity.** Furthermore this aspect of phase limitations can be used directly in genetics, when measuring number of copies produced in transcription of DNA(the so-called expression of genes). Speaking of the same wave(in this case: DNA sequence), by measuring phase limitations in the coding sequence, scientist may indirectly determine the frequency of copies produced during transcription process or the or the gene's expression in absolute numbers. What is a phase limitation in the context of DNA-sequence?

Genetic programming&ethical issues

Reading through language does not mean understanding the language. Creative approach to experimentally design constructs that would challenge DNA behavior in targeted organism raises ethical issues and may be regarded and morally compromised. However using the sequencing data in parallel with genetic programming language as attempt to understand molecular organization by using this elementary approach, first on a theoretical level, prior to launching targeted experimental protocols for in vitro or in vivo models, seem like natural pathway to foster the maneuvering abilities when it comes to the so called manipulation with the human genome.

3

FROM POLARIZATION TO GENE EXPRESSION

Can changes in membrane potential affect gene
expression?
Respiratory chain goes from mitochondrial
lumen to the mitochondrial intermembrane
space. It is a flow(transport) of electrons(negative
charges), cascade of oxidations and reductions
starting from the mitochondrial lumen, end-
ing in the mitochondrial lumen. They are pro-
totype of prokaryotes themselves(another concept). Mitochondria are simultaneously genera-
tors of energy. From inside, they are negatively
charged, like the rest of the organelles. Electrons
travel to the surface of the inner mitochondrial
membrane and then back to the lumen. Protons
go to the inter-membrane space. Mitochondria
at least partly utilizes proton (H+) pumps to
couple the electron transport? The answer is
trivial and has to do with water as the main con-
stituent of the cell and the energetic and polaric

black box or (rigid/amorphous) structure of the cell.

When action potential occurs, there is inward flow of cations, negative charges are pretty much fixed inside. But this inflow/outflow of cations have to be compensated. One of the compensators is Na/K ATPase driven by the flow of electrons in the inner mitochondrial membrane from the lumen(which is supposed to couple outflow of Na+) and the backflow of electrons from inner mitochondrial membrane to the mitochondrial lumen which is supposed to couple the inflow of K+. The mediator that articulates these changes in polarization in the cell is called ATP-energetic carrier.

What is the role of the sodium-proton exchanger on the mitochondrial membrane? Mitochondrial DNA is inherited maternally. Nevertheless electrons(negative charges) flow out from the mitochondria (an opposite concept compared with the cell membrane), while electrons in action potential flow out the cell. Prokaryotes do not have mitochondria and nucleus, eukaryotes do, its a discourse of a dogma. Mitochondria are considered important energetic source. They are just the source...

> ## Postulated definitions
>
> - Energy is the shift of potential.
> - Simultaneously means same change in polarization.
> - Polarity is affinity.
> - Charge is membership.
> - Cell currents induce magnetic fields perpendicular to their direction.
> - Proteins are conductors, lipids are isolators and carbohydrates are the energetic source.
> - DNA is inductor.
> - Membrane structures of the cell are capacitors.
> - Electromagnetic waves serve to signal in the cell, analogous to chemical signaling.
> - Metabolism is energetic turnover and "central dogma" is polarization turnover.
> - metabolism results from chemical signaling and "central dogma" from electromagnetic signaling.
> - Cell is LC-oscillator(inductor-capacitor network).

4

FUNCTIONAL ANALYSIS ON THE STRUCTURE OF NUCLEOTIDES

In this section the aim is to accentuate specific molecular patterns as to point out atoms and interactions of importance to the coding concept represented by the existence of the genetic code. The specific part, defining discrete difference in between nucleotides is the nitrogen base. Therefore further focuses is made on chemical phenotype and biological character of solely adenine, thymine, guanine and cytosine.

Adenine

Adenine consists of 4 ring nitrogen atoms and a single $-NH_2$ group linked to C^6. *Chemical phenotype:* corresponds to a decision-making nucleotide in the replication of the genetic information on every step, in terms of weather observed nucleotide sequence will be processed as replicating factor or an executing factor(e.g. extremes). *Biological Character:* Its abundance renders observed nucleotide sequences more conserved(prone to replication).

Guanine

Guanine consists of 4 ring nitrogen atoms, $-NH_2$ group linked to C^2 and O-atom linked to C^6. *Chemical phenotype:* corresponds to a translational nucleotide, regulatory in terms of what genetic information will be expressed on the cellular level and what is left to the conserving filter(see adenine). *Biological Character:* Its abundance discriminate further the executive importance of the genetic information for prompt use in the actual context.

Cytosine

Cytosine consists of 2 ring nitrogen atoms, $-NH_2$ group linked to C^4 and O-atom linked to C^2. *Chemical phenotype:* corresponds to guanine. It is executional nucleotide consisting of C^2 which links O by double covalent linkage and two N-atoms by single covalent bond. *Biological Character:* Nucleotides abundance terminates sequence-fate by fully conditioning the conservatory pathway and opting out from decision making role in the translational pathway. Both seem out of any significant cytosine-influence compared to any external factor or by chance.

Thymine

Guanine consists of 2 ring nitrogen atoms and a single $-CH_3$ group linked to C^5. *Chemical phenotype:* Thymine represents on-off switch for nucleotide-sequence processing in translational process. It increases the probability for pragmatic realization of the genetic information. *Biological Character:* Its abundance renders referring nucleotide sequence effective in the translational pathway. It couples adenine in DNA, so it complements its capacity to conserve the sequence.

5

MOLECULAR BIOLOGY&CELL BIOLOGY

Molecular biology has to do with molecules which are fixed or fairly static in space, rather than constantly moving and forming wave-looking changes in its molecular environment.

Cell biology postulates a changing balance, a constant movement to the point where "equilibrium" becomes unreasonable to mention.

The well established dilemma between mobile, charged particles and its molecular context(created from more or less fixed macro-molecules, their complexes or abundant molecules dominated by osmotic and other mechanic laws, e.g. H_2O) goes deep into mutually discriminating concept of structure and concept of function. The cell is their basic construct.

5.1 Some important consequences

Being mentioned as discriminating factors in the context of the cell, structure and

SOME IMPORTANT CONSEQUENCES 63

function(moving charges and fixed polarities&polarizations), are absolutely non-exclusive. However, given a specific process or part(extracted from the surrounding interaction/s and complex organization/s), they do not apply equally to the resolution of potential questions that arise. Simply put, while certain events are easy to prove by applying predominantly structural approach, other are provable with less obstacles upon using basically functional laws. Every postulate, properly defined, can be proved both ways, one of which much more difficult than the other.

The conceptual division of molecular and cell biology is usually given as a didactic instruction. Dialectically, these mechanisms are interspersed in cell processes and they depend on each other. Anyway, they are very useful in when proofreading new scientific ideas or theories, because of the reflections they offer on both size. The balance in these reflections gives prognosis for the ideas and their formulations.

5.2 The non-existence of housekeeping genes

Housekeeping genes are used for their (relatively) conserved&uniform expression across varied cell population to normalize the raw expression data for various genes in techniques that utilize PCR or similar protocols. For this propose $ddCt$ method is employed. One example is GAPDH(another one is Tpl2). The scale to perform such measurements is logarithmic meaning power exponent only in the exponential phase of the replication process is value of interest. Such value corresponds to the level of relative expression, when normalized to the activity on the housekeeping Gene. However housekeeping genes are a concept. As such their application must meet certain prerequisites, thereby creating more or less an idealistic artificial context. Firstly, their expression profile must have the same efficacy as the target gene(meaning the logarithmic base in the equation must be the same). That is hardly executable in real experiments. The series of equation below demonstrate that essentially any gene of interest could be

THE NON-EXISTENCE OF HOUSEKEEPING GENES 65

a housekeeping gene or regarded and used as
such or vice versa the existence of housekeeping
genes is relative and its postulation is completely
problematic given the true interest in the genes'
character. Their utilization in genetics is ana-
logue to introducing a balance in a changing
process.

Nowadays it is well established that various
genes not only determine cell fate, but also they
constitute the genetic programs which serve as
algorithms for regulating the activity and cross
regulation of coupled or related regions in the
genome. Therefore whenever researchers tackle
particular gene of interest, namely its expression under certain conditions, there has to exist
residual activity that can to a small or larger
extent predetermine expression profiling of specific gene or a set of genes. There may be other
rationals, as well, for introducing the concept of
residual expression activity which may become
significant due to its unique combination of regulatory sequences that can affect each gene. In
the field of cross regulation in between genes in

the genome the concept of relative expression value is irrelevant. For instance the expression of GAPDH-gene is often rendered identical, however it does not speak about its residual activity which may be robust, but is less likely to express uniformly when there is changing genome context or when different genes are examined.

What percentage of the expression value is specific and stick to the specific gene and what is the residual percentage that is constitutive product of regulatory sequences in the genome that differ with respect to that particular gene and to the changes in the regulatory sequences? Are there feedback loops? Is it positive or negative feedback?

$$dCt(A) = Ct(ID3inA) - Ct(GAPDinA)$$

$/$ $\qquad\qquad\qquad\qquad\qquad\qquad \log_x, (x \longrightarrow 2)$

$$\log_x (dCt(A)) = \log_x (Ct(ID3inA) - Ct(GAPDinA))$$

$$x^{\log_x (Ct(ID3inA)) - (Ct(GAPDinA))} = dCt(A) \qquad\qquad / \int dCt(A)$$

$$\int_{lowtreshold}^{hightreshold} (x^{\log_x (Ct(ID3inA)) - (Ct(GAPDinA))}) = \int_{lowtreshold}^{hightreshold} dCt(A)$$

$$Ct(A) = \log_x (Ct(ID3inA)) - (Ct(GAPDinA)) \times x^{\log_x (Ct(ID3inA)) - (Ct(GAPDinA)) - 1}$$

$$x \times Ct(A) = (cycles(A)) \times x^{(cycles(A))}$$

$, (x \longrightarrow 2)$

$$2 \times Ct(A) = (cycles(A)) \times 2^{(cycles(A))}$$

$for B :$

$$2 \times Ct(B) = (cycles(B)) \times 2^{(cycles(B))}$$

$relative\_ratio :$

$$2 \div 2 \times \frac{Ct(A)}{Ct(B)} = \frac{(cycles(A)) \times 2^{(cycles(A))}}{(cycles(B)) \times 2^{(cycles(B))}}$$

$$\equiv \frac{Ct(A)}{Ct(B)} = 2^{(cycles(A-B))} \times \frac{\log_2 Ct(ID3inA) - Ct(GAPDinA)}{\log_2 (Ct(ID3inB) - Ct(GAPDinB))}$$

$$\equiv \frac{Ct(A)}{Ct(B)} = 2^{(cycles(A-B))} \times \frac{\log_2(Ct(ID3inA))-(GAPDinA))}{\log_2(Ct(ID3inB))-(GAPDinB))}$$

$$\equiv \frac{Ct(A)}{Ct(B)} = 2^{(cycles(A-B))} \times \log_{Ct(ID3inB)-Ct(GAPDinB)} Ct(ID3inA)) - (GAPDinA)$$

$$(\log_{Ct(ID3inB)}(ID3inA)) \longrightarrow 1$$

5.3 Gene expression analysis: relative principle

Formulation

Introducing genetic leverage The term "leverage" is more commonly used in economics representing system of levers that increase the end-power of particular (profit-)mechanism[10].

In contemporary genetics the concept of leverage is more a statistical operation that increases positive predictive value of probable statements being made over the participation of certain gene in the etiopathogenesis of certain disease(in contrast to SNP and other natural variations in the DNA sequences that might be acquired or inherited).

Sexual reproduction is one example of genetic leverage by nature.

---

[10]usually the generation of profit by leveraging investments with borrowed money under relatively low interest rates.

1 Canonical criterium in gene expression analysis

Significance is derived from Taylor's formula given $f(x) = e^k$, where k represents a relative expression measure.

x $\approx$ characteristic valency for the gene of interest

x_0 $\approx$ characteristic valency for the reference(housekeeping) gene

$$f(x) = f(x_0) + (x - x_0)f'(x_0) + \frac{(x - x_0)^2}{2!}f''(x_0) + \dots + \frac{(x - x_0)^{n-1}}{(n-1)!}f^{(n-1)}(x_0) + Rn$$

Rn $\approx$ residual gene pool

$$Rn = \frac{(x - x_0)^n}{n!}f^{(n)}(\xi) \ , x_0 < \xi < x$$

$$e = 1 + \frac{1}{1!} + \frac{1}{2!} + \frac{1}{3!} + \frac{1}{4!} + \frac{1}{5!} + \frac{1}{6!} + \frac{1}{7!} + \frac{1}{8!} + \frac{1}{9!} + \frac{1}{10!} + \dots + \frac{1}{n!} + \dots$$

$$e = 2.7182819$$

1. Basic calculation
$$f(x) = e^3 + 6e^2 + 12e + Rn$$
$$f(x) = 97.039260236253753859259 + Rn$$

2. Threshold for residual gene pool

$$Rn = \frac{3^4}{4!}f^{(n)}(4.5)$$

Rn=3.375$(e^{9/2})''''$

$Rn = 199.3359375e^{1/2} = 328.649404495972118273921368333889$

Optimum number of cycles: 21.795829035592264654105231398998 $\qquad$ (p$\leq$ 0.05)

1 RT PCR plot

The following criteria is applied when a series of positive sequences[1] shapes the actual function of the sequences:

1. Cauchy's convergence test

2. D'Alembert's convergence test

$$u1 + u2 + u3 + \ldots + u_n + \ldots = \sum_{n=1}^{\infty} u_n$$

calculation:

$$\lim_{n \to \infty} \sqrt[n]{u_n} = L$$

$$Cauchy|criterium$$

$$\lim_{n \to \infty} \frac{u_{n+1}}{u_n} = L$$

$$D'Alembert|criterium$$

If

- $L<1$, *series is convergent*

- $L>1$, *series is divergent*

- $L=1$, *series is undetermined*

---

[1]mathematical term

Background in theory

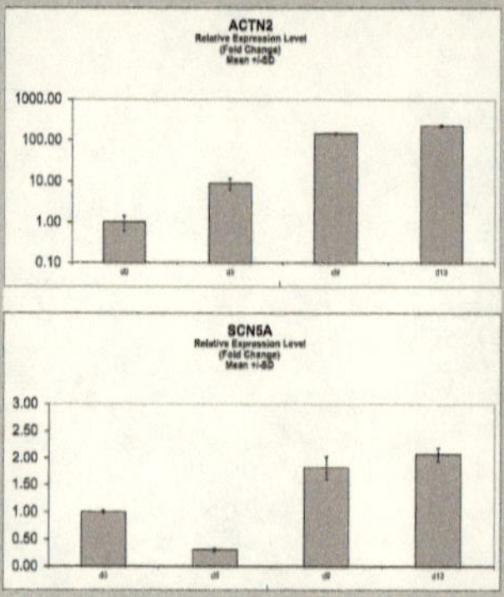

Each column in the plots represents a separate sequence

1.1 Graphics

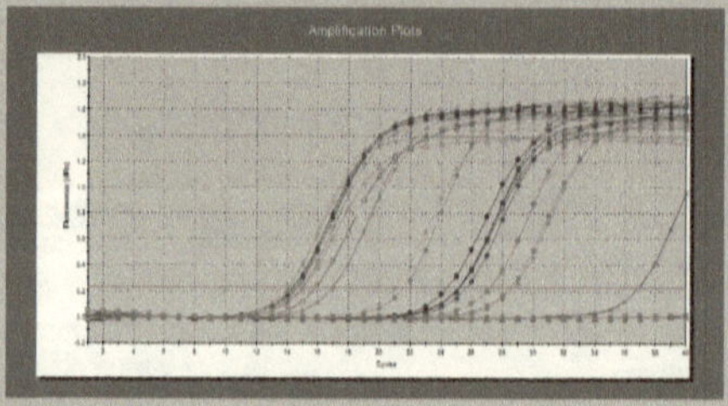

Functional analysis

3

1.2 ROLLE'S THEOREM

If $f(x)$:

1. is continuous and differentiable on the closed interval $[a, b]$

2. attains equal values at two distinct points(e.g. $f(a) = f(b) = 0$ or $f(a) = f(b)$),

3. there has to be a stationary point($x = c$), where the first derivative equals 0;

$$f'(c) = 0, a < c < b$$

1.3 Mean value theorem[2]

If $f(x)$:

1. is continuous and differentiable on the closed interval $[a, b]$

2. for every point, there is a final derivation

3. there has to be a stationary point($x = c$), where:

$$\frac{f(b) - f(a)}{b - a} = f'(c), a < c < b$$

or

$$\frac{f(x_0 + \xi) - f(x_0)}{\Delta x} = f(\xi), x_0 < \xi < x_0 + \xi$$

---

[2]first described by Parameshvara (1370–1460)

4   Background in theory

2 Numerical strategy

$$Ph = ln(2) = \frac{lg2}{lge} = 0.69314718055994530941723212145818$$

for dRn=0.2325.(refer to the table below)

2.1 Confounding variance

2.1.1 ACTN2

L0=1
L5=2.95
L9=5.24
L13=3.86

The output suggests consistent and convergent data at the chosen time-points.

2.1.2 SCN5A

L0-4=0.2958
L5-8=6.1125
L9-12=1.1381
L13=/

For observing the ANOVA-statistical output, refer to the **'Results'** page.

ELABORATED BY: KLIMENT SANDJAKOSKI, MD
TMR PROGRAM

The concept of polarization pressure

Polarization and pressure, quoted separately, are well known terms in physics that define very different occurrences: charged particle is often denominated as polarized, whereas pressure is physical variable that is produced whenever a force is applied to a unit area of surface. Polarization implies electrostatic force that acts by the Couloumb's law.

Cell membrane is polarized. Other membrane structures in the cell are polarized as well. Action potential represents an event on the cell membrane followed by moving charges that resolves with negative feedback, driving ion-concentrations into initial state and recovering resting membrane potential. Two properties that remain during those dynamic processes are: semi-permeability of the cell membrane and the (macro)molecular structure of the cell. Those properties, however, transiently change during action potential or some signaling processes, but their initial state is being quickly recovered. Nevertheless, the negative

GENE EXPRESSION ANALYSIS: RELATIVE PRINCIPLES

(macro)molecular composition of the cell represents the source of electrostatic force rendering cell surface negative on the inside, whereas surface area is represented by the semipermeable cell membrane. Hereby, resultant electrostatic force vector and surface create the concept of polarization pressure. What is polarization pressure? Is it the bidirectional semipermeable state of the cell membrane? Pressure is a scalar quantity. Anyway, polarization pressure has direction, so it is a vector. It represents summarized gradient that goes beyond semi-permeability of the cell membrane and impacts cell polarity, creating context for molecular(including genome) events that disciplines intracellular processes to the central dogma of biology. Relaxation of the polarization pressure creates molecular context for external input into intracellular processes through signaling or passivation of those molecular processes.

In summary:

1. polarization pressure could increase or decrease,

2. it always has positive value and

3. action potential resets or purges the cell from excessive or insufficient polarization pressure

What renders this definition of polarization pressure, as summary gradient, resulting from negative intracellular composition and semipermeable cell membrane, real? Why is this concept not applicable to various polarities of (macro)molecules and their complexes in the cell?

Firstly, it is obvious that negative polarity is not uniformly distributed at various points in the cellular content. Therefore, it is also well known that electrochemical gradients exist between different parts, structures and compartments in the cell. Whenever polarization comes in place, we speak of context that is robust, irrespective of structural, energetic or polar changes in the molecular biology of the cell. Furthermore polarization is discrete and independent from intracellular events that affect polarity, energy or structure, consequently, polarization pressure concept is not necessarily correlated to intracellular properties when regarded separately, but

GENE EXPRESSION ANALYSIS: RELATIVE PRINCIPLE 77

rather it shapes the buffer capacity for processing signals which may impact the cell to a larger or lesser extent depending on the concordance of the actual polarization pressure and its vector with the signal.

In search of an ultimate regulator of gene expression

Introducing genomics..1990s

1. One gene at a time

2. Revolution of genomics

3. time-dependence in the concepts of gene expression and membrane-potential-changes

4. Ca as key-particle of the external impact on the cell

5. Univalent cations, together, create the context of cellular behavior

Enzymes that add or delete acetyl, methyl, and phosphate groups must be in a balance that controls which genes are expressed and which are silenced. [11]

- Agonism to ion channels or related receptors increases frequency, but not the amplitude of [Ca] oscilations

---

[11]Lewis R, 2009, Human Genetics (9th ed.), DNA and Chromosomes, 206p

- Increase in cytoplasmic [Ca] produces increase in [Ca] in the cell nucleus;

- A rigid structure in the cell that resists to the polarization pressure established by membrane potential and challenged by its changes?

- Increase in cytoplasmic [Ca] produces increase in [Ca] in the cell nucleus

Changing context of ions in the cytoplasm and nucleus

- Cytoplasmic concentration

- Nuclear-orientation and direction

Cytoplasmic vs. nuclear Ca

- fine-tuning gene expression

- Nuclear Ca

- chromatin remodeling as an on/off switch to transcription

Ionic transients impact

1. Recombination

- signaling component-breakage and rejoining of DNA strands

- chromatide adhesion-cycling component(Na, K,..)

GENE EXPRESSION ANALYSIS: RELATIVE PRINCIPLE79

Gene expression

- expression-cycling component(Na, K,..)

- gene selection-signaling component

Changing concepts in signaling mechanisms
Canonical vs. noncanonical

Calcium

1. Cytoplasm: role of cytoplasmic proteins in terms of Ca

2. Nucleus: role of changing Ca inflow/outflow

   Na, K, (monovalent cations)

- Cytoplasm&Nucleus: role of monovalent ion-exchange between nucleus, cytoplasm and extra-cellular environment

Questions that arise

- Impact on mendelian genetics?

- Impact on recombinations as a diversity creating and experimental concept?

- Role in the immune system?

- Role in neuroinflammation?

- CaMKII [4][5][6][7]

- Calcineurin [8]

Calcineurin is active whenever Na/K ATPase is active?

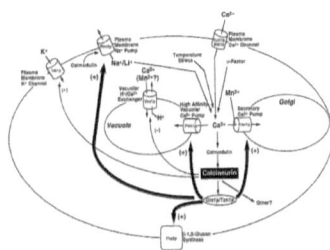

[The reason for choosing these molecules is recent evidence that shows how they regulate Na, K and Ca channels.]

6

FEEDBACK: GENE REGULATION&MEMBRANE POTENTIAL

Do gene regulation and membrane potential reflect each other? If so, is this (co)relation robust? Finally might we regard membrane potential as a function of the genomic state. Genome is self-renewable and membrane potential is conserved. Those two categories are both activated when it comes to cell death(induced or apoptotic). Furthermore are they interdependent? If so, to what extent?

Those and other questions stems from the old dilemma about chicken and egg. Today we know that cell membrane is not simply a barrier but also represents a potent, selective, enzymatic, metabolic, signaling, transport system which utilizes energy to maintain its functional state.It manages cell communication in between cells, but also to the extracellular space.

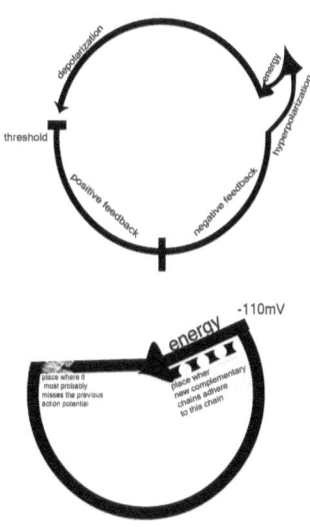

6.1 Positive feedback-minor changes in membrane potential

6.2 Negative feedback-Action potential

Action potential is most likely to represent a reset button for polarization changes at the level of cell membrane. It's been known for a long time now that it can be triggered by chemical signal or by surpassing threshold for certain voltage drop. The widely entertained physiological

NEGATIVE FEEDBACK-ACTION POTENTIAL   83

principle "all or nothing" will be important in further discussion. Ions follow so called electrochemical gradients whenever opening of specific ion channels takes place. It is noteworthy as depolarization and repolarization phase act as coupled processes.

However cell membrane as any other biological membrane in an in vivo setting is exposed to ever changing polarization gradients, pressure gradients as well as concentration gradients. It is semipermeable barrier. Voltage barrier is byproduct of largely negatively charged, relatively static macromolecular complexes that attract positive charges from marginal spaces, which as a result renders cell membrane negatively charged on the inside. If any polarized particle passes the above mentioned barrier, it will have no significant influence on the membrane voltage, meaning time frame and/or the amount of polarized particles that crosses one specific protein-channel is of key importance. Simply put, intracellular polarization state vary even when voltage of the cell membrane remains the same. Until action potential takes place. It resets not only voltage across cell membrane, but polarization context of intracellular content. It reduces polarization pressure(tension in between electrical and concentration gradient) towards zero. From genetics perspective, action potential washes off all changes in the protoplasm that had not been materialized as signal to DNA-RNA-protein axis. Anyway, action potentials, themselves, do change the quality of ions map in the cell(K+ ions are lacking and Na+ are overwhelmingly present inside the cell). Concentration of Ca2+ ions change as the other electrolyte concentrations do change. Active transport is carried out through ATP-ases, ion-exchangers and other specialized channel proteins to recover the qualitative representation of designated ions to its original state.

Thus sensitivity of the action potential is inherent cell feature, whereas its specificity is guided by the ability to communicate other cells and the natural limits to its inclusiveness.

NEGATIVE FEEDBACK-ACTION POTENTIAL   85

Corresponding phenomena: Long term potentiation(LTP) Long term depression(LTD)&Short term potentiation(STP)&short term depression(STD) in learning and cardiac memory

The continuous electrical activity is maintained largely by colloquial cross sectional *reset* points that are mainly represented on the level of synapses. The electrochemical signal on a synapse level converts the electrical oscillatory signal into chemical signal that follows saturation curve. This conversion occurs in varying conditions and with different efficacy. This renders synaptic transmission a time- and space-dependent reset point.

7

CELL CURRENTS

Abstract*

There are currents in the cell...

1. On the inside of the porous 3D capacitor called cell membrane. Currents induce magnetic fields rendering the cell-organization depend on its structure&function simultaneously.

2. Simultaneous is defined as same change in polarization.[12]

---

[12]My definition for the purpose of this book

EXISTENCE OF CELL CURRENTS

7.1 Existence of Cell Currents

Advocating cell currents is expected to contradict the existing polarity in different compartments inside the cell. Those polarities are used to determine direction wise the flow, transport and migration of specific molecules, metabolites etc. The energy is usually derived from molecular energy carriers as ATP, GTP etc. One example for this are microtubules, but other elements of the cytoskeleton are involved as well.

On the other hand it is fairly impossible to explain the readiness of the cell for automated response in specific manner to external also specific chemical, electrical or mechanical inputs when molecules or other micro-particles are applied at a specific location on the cell surface. One emerging field, namely cell signaling imposes well established mechanisms where molecular cascades are employed to explain their robustness, however there is no conclusive evidence, at least, not in practical terms rendering signaling cascades irreversible. External stimulus provokes a signal, but its effect remains robust from an outsider perspective. Molecular cascades, regarded solely, are exhibiting consequences irrespective to the membrane structures where they are located(position in the cell). Their responsiveness is determined by the functional capacity of the cell, so in terms of continuous flow they never reach a dead end, whereas signaling cascades are discrete, preconditioned by stimulus and the outcome is not proportional or even the same quality as the initiating event. The notion of cell current derives from the idea that molecular evolution [13] is relevant only in the context of existing generalized "state of cell", creating a context for exhibiting molecular features. In other words, cell currents are actively conditioning the molecular process, thereby executing its fate and determining the outcome.

Cell currents are postulated in this chapter by their consequences, while presented as an elephant in a dark glasshouse-the smell is there, as we try to find the switch.

---

[13]Molecular Biology Of The Cell, Alberts et al.

INTRODUCING QUATERNARY NUMBERS TO NITROGEN BASES

7.2 Introducing quaternary numbers to nitrogen bases

Quaternary numeral system is naturally considered when analyzing four basic constituents of the DNA-chain. Denominations for $N_4=0, 1, 2, 3$ are C, A, T(U) and G, accordingly:

$C = 0; A = 1; T(U) = 2; G = 3$

The parallel is to be made as 0 resembles the grounded phase in three-phase electric power and each one of the remaining quaternary numbers represent the other three phases.

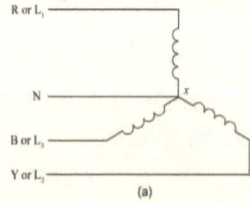
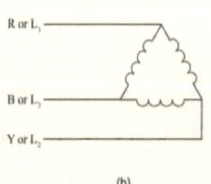

(a)   (b)

CELL CURRENTS

Genetic code in base − 4 numeral system

UUU 3132	UGC 3212	CAA 23	ACG 1213	GCU 10232
UUC 3130	UGA 3213	CAG 31	AAU 1300	GCC 10230
UUA 3131	UGG 3221	CGU 200	AAC 1232	GCA 10231
UUG 3133	CUU 112	CGC 132	AAA 1233	GCG 10233
UCU 3022	CUC 110	CGA 133	AAG 1301	GAU 10320
UCC 3020	CUA 111	CGG 201	AGU 2010	GAC 10312
UCA 3021	CUG 113	AUU 1322	AGC 2002	GAA 10313
UCG 3023	CCU 2	AUC 1323	AGA 2003	GAG 10321
UAU 3110	CCC 0	AUA 1321	AGG 2011	GGU 11030
UAC 3102	CCA 1	AUG 1323	GUU 11002	GGC 11022
UAA 3103	CCG 3	ACU 1212	GUC 11000	GGA 11023
UAG 3111	CAU 30	ACC 1210	GUA 11001	GGG 11031
UGU 3220	CAC 22	ACA 1211	GUG 11003	

INTRODUCING QUATERNARY NUMBERS TO NITROGEN BASES

Quaternary algorithm for the DNA syntax

GTCAGCTTCGCAACGTCAGCTTCGCAGC CTGCACGTCCC
TGCAGTCGAAGCGTTGCAGTCGAAGCGTCGTG GTGCAGGG

10320130220301103201302203013010230103200
230132031130322301320311303203231032301333

CCCCCCCCCCCCCCAAAAAAATTTTTTT
GGGGGGGGGGGGGGGTTTTTTT

00000000000000222222211111111
3333333333331111111112222222

CGCGCGCGCGCGCGAAAAAAAAAAAAAA   CACACA
GCGCGCGCGCGCGCTTTTTTTTTTTTTT   TGTGTGT

0303030303030311111111111111   1010101
3030303030303022222222222222   2323232

CACACAC   TGGCTGCTACCCG
GTGTGTG   CCCAAGGTGGCC

0101010   2330230210003
3232323   1003103123330

Base$_4$(10320130220301103201302203013010230103000) =
=102331020001331020331222333000000000000000000000
00000000000000000000

Base$_4$(2301320311303223013203113032032310323011333) =
=2221010300120333131310032320000000000000000000000
0000000000000000000

Base$_4$(0000000000000022222221111111)=11003120003212012331013

Base$_4$(33333333333333111111112222222)=
=2230111010302201102312323020000000000000000000000

Base$_4$(030303030303031111111111111111)=
=332222210221221030223120320000000000000000000000
Base$_4$(30303030303030222222222222222)=
=213022212322001233323002200000000000000000000000

Base$_4$(1010101)=3312212311
Base$_4$(2323232)=20313030200

Base$_4$(0101010)=120222102
Base$_4$(3232323)=30111021003

Base$_4$(2330230210003)=201322030202303313103

Base$_4$(1003103123330)=32212031212233132002

Complete Base$_4$ number is: 33213021002312201230130320200000000
0000000000000000000000000000000000000000000000000000000000
0000000000000000000000000000000000000000000000000000000000
0000000000000000000000000000000000000000000000000000000000

7.3 Cell depolarization&repolarization plot

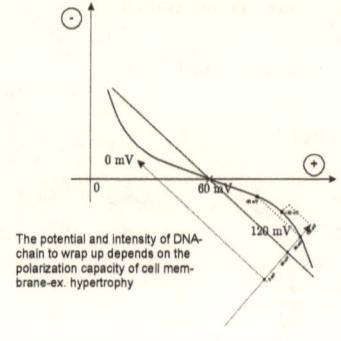

0 mV

0   60 mV

120 mV

The potential and intensity of DNA-chain to wrap up depends on the polarization capacity of cell membrane-ex. hypertrophy

$y=bxl^n$

n is changing parallel to the level of condensation

$y=1/x$

y-frequency of action potentials

$lxl^n=1/x$

$xlxl^n=1$

n-odd number

$x=1$

94   CELL CURRENTS

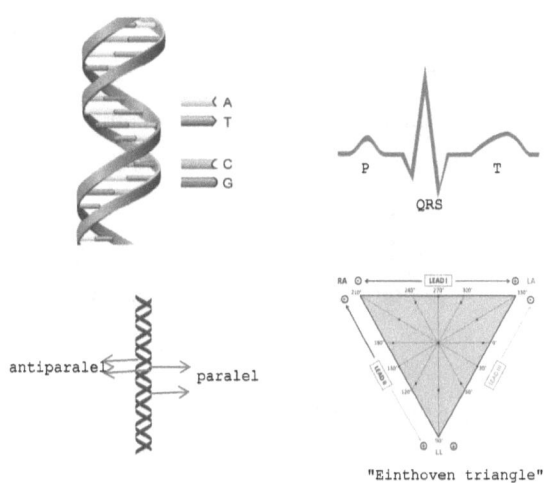

antiparalel   paralel

"Einthoven triangle"

paralelogram

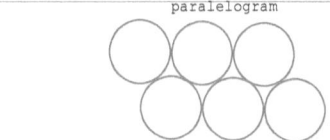

channels on ECG

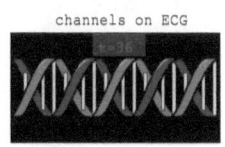

"war industry"

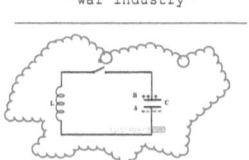

"pharmaceutical industry"

инерција

F=m*a   m*v1=-m*v2

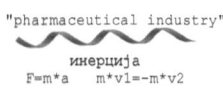

7.4 Cell currents in signaling

Signaling is considered primarily molecular event carrying signal through the cell, resulting in specific change which is proportional and adequate given a certain predisposed cell state. It utilizes specific molecules and produce specific changes in their physiological behavior. It couples energy and charge, designated at predetermined molecular pathway.

Can signalling be reversed?

One particularly hard path to follow, it is trying to mimic the cascade reaction in search for the opposite solution. Signaling pathways are unique, fast efficacious, partly due to the involvement of moving polarities and the orientation is another mostly complicated issue.

Anyway, observing it from the perspective of existing cell currents this concept is not unattainable. Until recently signaling was fundamentally divided on two arbitrary categories:canonical and non canonical. Nowadays researchers realize that the two branches are interspersed and interdependent in real time

96   CELL CURRENTS

that any division becomes unreasonable.

How can cell sense discrete molecular polarity signals without producing significant change in membrane polarization?

The popular response to the key question above is the well entertained concept known as chemical affinity of signaling molecules to energy-coupled receptor transmembrane proteins. But this does not seem to answer fully the question. The postulated answer in this book would be: Cell Currents! Only electrical current could perform sensing of such microscopic stimulus and execute proportional output. Transformations of energy occur on the way and that is not what is challenged in this text, furthermore while cell currents are there, they need to be characterized.

A context where charges are either moving, trembling or surrounded by apolar (macro)molecular units on one side and static polarities on the other.

CELL CURRENTS IN SIGNALING　　　97

Relativistic perspective on cell structure

Given a perspective of an external observer cellular membranes looks as usually described-planar structure having negative charges on the inside, building capacity on the hydrophobic layer in the membrane, whereas nucleic material is organized in helix(more or less condensed) having spiral, helical structure. Thus a **centripetal** perspective on the cell structures represents only one side of the story.

Advocating an opposite position, starting at the DNA-level, no special image processing is required to represent more or less β sheet(zik-zak)-like structure that orients and reorients surrounding cellular environment **centrifugally**. As a result the cell is intrinsically divided on resultant polarities that create a context for cell polarity, another important topic in molecular and cell biology. Important mechanisms involving cytoskeleton are explained by the notion about "cell polarity".

Cell currents are set be only tool to reassemble the logical collision of the conflicting discourses.

7.5 In vivo molecular characterization

Categorization of organic macromolecules intro-
duces considerable order to their in vivo coex-
istence, in terms of creating specific molecular
context or in terms of their mutual interactions.
In modern day science four basic groups are well
established: carbohydrates, lipids, proteins and
nucleic acids. All four groups have distinctive
biochemical features regardless of its potential to
co-opt in diverse molecular processes. However,
diverting their specific role in cellular context
almost always comes at a certain cost. One
example are nuclear proteins(e.g. histones) that
participate in the proper molecular organization
of DNA, they can be passed on to the next generations and together with other sorts of organic
macromolecules consist the epigenetic molecular complex. They become relevant in many
contemporary research and clinical efforts to further randomize in vitro fertilization(IVF) as well
as clinical genetics per se. Nevertheless nucleic
acids are a starting point when regarding any
genetics' problem of clinical relevance or any

kind of research question.

The rationale to study nucleic acids by its unique molecular characteristics derives from a common struggle to put nucleic robustness and stability in relation with their capacity to sense as to respond on the functional role of the remaining groups of macromolecules.

By caring comprehensive recapitulation so far known to attribute structural and functional role models for their molecular behavior in particular cellular context, few collective properties seem very much worth considering. Thinking from perspective of purely chemical setting, lipids characterizes ester linkages, proteins the amide linkages and polysaccharides the glycosidic linkages, whereas polynucleotides have phosphodiester linkages. Those linkages set common ground for further investigating the collective behavior of the four groups.

A contemporary view on homoeostasis

Although translation is regarded as unidirectional process and so the proteins, as regulators in chief, in the context of maintaining particular physiological state, these assumptions may not hold true or simply put are utterly incompatible when considered the facts and principles entertained in modern day biomedical science. Proteins, regarded solely, do not leave options when applied to a specific process in the cell. They themselves derive from a larger pool of coding sequences in m/iRNA and backward to the DNA-gene. Splicing and other interventions take place on that molecular pathway. Furthermore, proteins(and their activity and/or structural participation in a specific molecular context) have the ability to communicate and be influenced by cell signaling processes as well as be preconditioned by access to energy. One good example are ionotropic and metabotropic receptors in cells, basically proteins that are preconditioned to execute their function by gradients in polarization(ionotropic) or adaptive changes in terms of static level that creates difference in polarity by changing molecular byproducts.

IN VIVO MOLECULAR CHARACTERIZATION   101

Therefore, homeostasis is erroneously represented more likely to represent a philosophical
denomination for "balanced" state of the cell
or predefined and well controlled equilibrium.
The first definition excludes constant changes
that are unobtrusive in every living organism,
whereas the second one corrupts future understanding for new things that will significantly
change our perspective for the ongoing discus-
sions. Hereby, defining homeostasis remains pure
utopia.

Energetic articulations

Energy in the context of cell metabolism may
be regarded as free flow of negative charges.
A chaotic matter charged for potentiation or
exhaustion of coupled reactions and cascades(in
molecular biology), and directional flow of
ions(in cell biology). If regarded so, carbohydrates are imposed as most suitable energetic
source, assuming that energy in a biological
context is almost always supplied by breaking

down as many highly energetic chemical linkages. Carbohydrate metabolism is almost continuous process that could be stopped, reversed
of fostered at any point, circumventable, fast
and convenient, rendering carbohydrates a per-
fect candidate for caring out the conditioning of
a desired energetic state.

Marginal burst

The long process of molecular evolution has
left very few details unresolved. From a standpoint of a modern day scientist-some processes
occur almost continuously while others are so
rare and short lived. The same holds true for
the "central" molecules that get involved in
those processes. However their utilization rate
is not always proportional to the level of abundance in an *in vivo* biological system. In certain
functional compartments they become structural constituents in the higher organization of
complex organic macromolecules, thereby forming biological complexes and contributing in
its features. Similarly they take part in biochemical complexes that belong to more than

one group in our basic categorization in four groups(lipoproteins, glycolipids, etc). It is only partly due to the fact that some of their characteristics are not replaceable with other molecules, another aspect is their capacity to back up concepts which are primarily invoked for specific mechanism so they serve as reasonable second hand option.

Lipids entertain a fair bit of the above mentioned features. They have the ability to reversely concentrate metabolic energy surpluses. Considering their inert behavior(when it comes to energy supply for metabolic, transportive and other activities) within the cell together with the hydrophobic, nonpolar character in a purely chemical setting, they position on opposite ends with carbohydrates, given a certain differentiation vector in the context of the curious field of molecular evolution.

Phosphodiester bond

Nucleic acids have this common chemical linkage, that explains a whole set of major function to actively code for the entire proteome, and

CELL CURRENTS

pass it down in to the next generations.[14] Higher molecular structures and conformations, in the case of nucleic acids, become precise regulators of their level of activity. They respond to physiological and biochemical signals in a sensitive and specific manner, representing accumulated knowledge on cell character.

Molecular evolution

Contemporary science regards molecular evolution limited to DNA-RNA-protein axis. In the same context existing redundancy is regarded as differential expression of redundant genes either spatially or temporally. From a perspective of a living cell, natural selection is performed on a genetic level, weather the abundance of environmental resources(carbohydrates and lipids) is part of the so called neutral evolution. Science tools seem insufficient to tackle questions such as which enzymes were made first or what were their products in the early days of primitive biological systems where adaptation to a living conditions plays major role.

---

[14]There are certain proteins(histones) that daughter cells inherit automatically, rendering the field of epigenetics notably important and relevant.

8

INHERITANCE PATTERNS

8.1 Mendelian&non Mendelian

Traditionally two major types of inheritance patterns are entertained in the contemporary biomedical science:

• Mendelian inheritance segregates alleles whereby dominant are expressed in heterozigous. Three Mendelian laws are followed in such inheritance pattern:

> segregation
>
> independent assortment
>
> dominance

• non Mendelian inheritance consider inheriting two or multiple alleles that does not comply to Mendelian patterning rules.

Mendelian inheritance represents a reflection of biological, behavioral as well as evolutionary models overloaded in common dogmas in various natural sciences. It represents a force-driven process, exposing traits that are meant to be better adjusted to the "genetic context" compared with the opposite allele. Noteworthy, one precondition is the existence of competitive allele coding alternative traits. However this is not the case in all genome.

Certain phenotypes present with traits that differ significantly from the homozigous state. Furthermore, their variance exceed the number of alternative traits calculated by the dominance law. Conclusively, competition is not always welcomed in the inheritance field. Therefore, the term-non Mendelian inheritance is utilized to join remaining inheritance patterns that are also represented by alternative traits, but do not follow Mendelian rules.

Is it manifesting certain inheritance mechanism we know nothing about or it is endless group of inheritance patterns having nothing in common, increasing genetic entropy characterized by their spontaneous expressiveness or is it stochastic?

By applying the relativity principle in the world of inheritance, a whole new theory comes to mind. Mendelian laws are applicable only for alleles that depend of negligible other transcripts(genes products) on their pathway through

the cellular meshwork to the realization of particular trait. **In other words, Mendelian laws are applicable only in those genes where their segregation depends solely on the sequence of the gene, not the related, surrounding of interfering nucleotide sequence from other coding or non coding sequences of DNA.**

Mendelian rules deriving from the Mendelian laws of inheritance are useful in modern day genetics they are applicable on large scale of genes, genome wide. However they do not provide sufficient explanation to whole set of very important everyday questions arising from central genetic events, mechanisms and wide range of regulatory and repair processes in the genome.

Non Mendelian inheritance implies:
- extranuclear inheritance
- gene conversions
- infectious heredity
- genomic imprinting
- mosaicism
- trinucleotide repeat disorders

Both Mendelian and non Mendelian inheritance patterns are provisional denominators of what has been already acknowledged in the field of genetics during the past half-century.

Although they are set to co-opt in a rule &exception manner, inheritance has to be way more complex as a process...and it is more complex process.

8.2 New *complex inheritance*

Complex inheritance principles

1. *there is order in between genes in the coding process*

2. *gene expression is first priority when setting up the order*

3. *the order is encoded by DNA-sequence **and** in the so called junk DNA, both being equally important*

4. *transcripts, proteins, infectious agents are regarded as environmental(outside) factors*

   Relativity principle imposes that altered sequence in a single gene would less likely create difference in segregation plot, when it comes to dominance. Therefore, as a result there will be more likely a Mendelian order. However, when large group of genes are coordinated in their inheritance patterns alteration in any of the gene sequences of the participating genes can

significantly and sometimes drastically alter the likelihood for executing the specific trait(altered dominance pattern).

Moreover, when single gene alleles segregates by a specific pattern, this could be easily altered when extensive changes of nucleotide sequence within or in the near "gene environment" occur.

9

CURIOUS DESIGN IN THE FIELD OF DRUG DISCOVERY

Pivotal role in understanding the DNA sequences as well as their meaning in transcription and their translational power, as a whole new concept, plays creative drug design and the entire field of drug discovery. By utilizing coding vocabulary, scientist may *know*(not only progress disclosing further members of more less accidental discoveries in the recent history of clinical medicine), but unmistakably design drugs that are appropriate to their specific therapeutic goals, at the desired rate.

By using transformation(rather than progressive) methodology, qualitatively new types of molecules and new groups, families and classes of drugs can be created.

9.1 Original examples

The concept of targeted creation of new pharmaceutical drugs that would act on specific

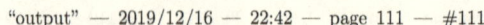

ORIGINAL EXAMPLES   111

molecular object sprouts from the idea that the physiological state of the cell as a whole as well as the discrete molecular processes that define certain state are regulated from a coded message realized in the processes of transcription and translation.

Methylxanthines

Methylxanthines interfere with inhibitors of DNA-gyrase on the level of RNA.

Biotin

Corrected formula

Metamizole

Two matimazole moleculaes coupled through Ca ion, form enhanced version of the existing pharmacy.

Alprazolamum

One sodium ion is missing in the alprazolam chemical formula, right next to the chlorine-atom

112   CURIOUS DESIGN IN THE FIELD OF DRUG DISCOVERY

Figure 9.1: Two metamizole molecules bridged together by single Ca-atom

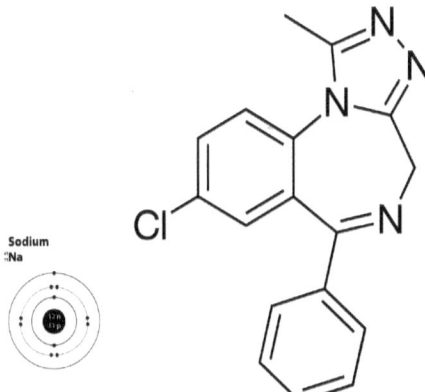

Figure 9.2: Alprazolam with approaching single Na-atom

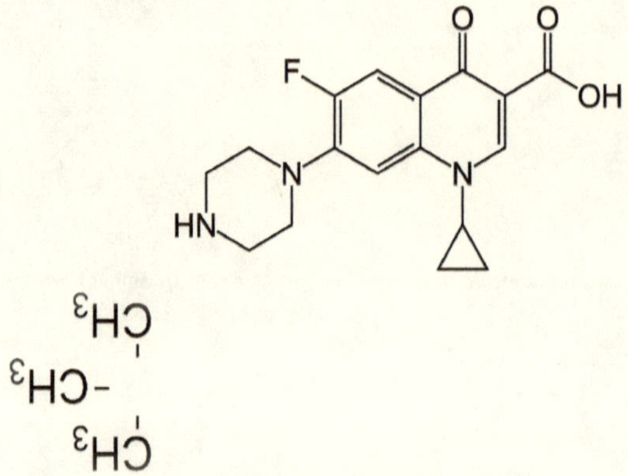

Figure 9.3: Ciprofloxacin with additional three methyl groups

Ciprofloxacin

In the chemical formula of ciprofloxacin, three methyl groups should bind fluorine-atom.

10

RADIOACTIVE PHYLOGENETICS

10.1 Reflections of the atomic structure

Orbitals

There are 6 different types of atomic subshells deriving from the nine recognized shells, surrounding the atomic nucleus in atoms. The six subshell-types and their reflections are listed below:

1. *s-subshell(max: 2e⁻): primitive concept of coupled interactions*

2. *p-subshell(max: 6e⁻): concept of conflict*

3. *d-subshell(max: 10e⁻): concept of couple*

4. *f-subshell(max: 14e⁻): concept of offspring*

1	K	=	2	$1s^2$
2	L	=	8	$2s^2\ 2p^6$
3	M	=	18	$3s^2\ 3p^6\ 3d^{10}$
4	N	=	32	$4s^2\ 4p^6\ 4d^{10}\ 4f^{14}$
5	O	=	32 *50*	$5s^2\ 5p^6\ 5d^{10}\ 5f^{14}5g^{18}$
6	P	=	18 *32*	$6s^2\ 6p^6\ 6d^{10}6f^{14}$
7	Q	=	8 *18*	$7s^2\ 7p^6\ 7d^{10}$
8	R	=	2 *8*	$8s^2 8p^6$
9	S	=	2	$9s^2$

120 - 170

5. ***g**-subshell(max: 18e$^-$): concept of coupled individual*

6. ***h**-subshell(max: 22e$^-$): concept of single individual*

10.2 Periodic table, elements, subshells, orbitals and Fourier transform

Subshells, on the orbitals' level, are applied a concept represented by the Nyquist rate for sampling of the proton content in the nucleus. Their orbitals are filled once each and every orbital has at least one electron of maximum two. Then they raise the number of their electrons to max. 2.

For instance, if an atomic nucleus contains seven protons there will be 10 electron-positions in all shells, seven of which are filled with electrons. This means that two s-subshells are fulfilled and p-subshell is half-fulfilled(there are three electrons distributed in the three orbitals of a p-subshell, one electron in each p-orbital). Furthermore the two electrons in each orbital always have opposite spins: $+1/2$ & $-1/2$.

It may be postulated that orbitals serve as a sampling rate for registering proton events which are designated as electrons. Electrons have varying energy depending on their position in the atomic shells and subshells, so they could represent reflections of proton events that exit the atomic nucleus.

However, regardless of the order in which empty electron positions are fulfilled in orbitals, subshells are fulfilled in a discrete manner-one by one in predesignated order deriving from their free energetic capacity. Namely p-subshells begin to fulfill once s-subhell is fulfilled and p-subshell begins to fulfill once subshells of the previous energetic shells are fulfilled(d-subshells, f-subshells, etc.). Comprehensively, the next s-subshells are energetically closer to p-subshells of the preceding shell than d-,f- and higher subshells of the preceding shell. Consequently each subshell represent discrete frequency of a possibly continuous electromagnetic wave that exists in limited time-interval, different from the frequency of other subshells. Conclusively

PERIODIC TABLE, ELEMENTS, SUBSHELLS, ORBITALS AND FOURIER TRANSF

fourier-transorm can not be applied on noncoherent waves, apart from using a modification of fourier-transform for discrete intervals. However there are major technical challenges to achieve compromising value of time interval because different frequencies have varying duration. In this context the period between s-subshells in consecutive shells progressively increases by increasing amount. Running Fourier transform on such continuous non coherent wave-representing the quality of electron-cloud around the atomic nucleus can therefore speak of the structure of atomic nucleus but its almost impossible to track its functional state in terms of intranuclear events that reflect on energetic structure of electron cloud.

The above mentioned concept is important due to the existence of natural tendency on the atomic level to fulfill or discard partially fulfilled atomic subshells. This is apparent in chemical bonds of different kind where complementary atoms conform chemical bond. The stability it brings to the atoms is a represent of complementary functional state of both atomic nuclei whereas bond's fragility is due to incompatibility of the underlying atomic shells and subshells of both nuclei(energetic inequities of complementary subshells). *Simply put, the higher the energy to disrupt certain chemical bond(more stable the bond) the more complementary proton events are in the participating atomic nuclei. The more unstable a chemical bond is, the bigger the discrepancy is in structure of their electron cloud.* Assuming the robust nature of proton behavior in the nucleus and by calculating the energy of certain chemical bond we may discriminate for the proton events in the larger nucleus.

Spontaneous radioactivity

Concept of spontaneous radioactivity

10.3 Reflections on the genome

Radioactivity has socio-organic reflections that result in predictable phylogenetic outcomes. The basic law defining principles of nuclear decay is:

$N = N_0 2^{-t/t_{1/2}}$

N-number of genes that affect expression of a

REFLECTIONS ON THE GENOME 119

gene of interest(gene which is responsible for certain abnormal condition)

N_0–number of all genes in the human genome

t–maximum number of steps that may produce significant effect on gene's expression

$t - 1/2 - $–number of steps required to change target's gene activity by 50%

This remains fully valid for pedigrees in every sexual reproduction of organisms.

1. α decay: heterosexual relationship

2. β^+ decay: bisexual relationship processing

3. β^- decay: homosexual relationship processing

4. γ decay: friendly relationship

RADIOACTIVE PHYLOGENETICS

(a) $^3_2He->stable$

(b) β-decay: $^3_1H->^3_2He+\upsilon$

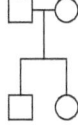

(c) $^4_2He-stable$

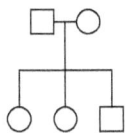 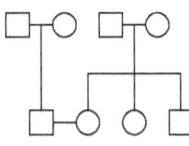

(d) Proton emission: $^5_3Li->^4_2He+p^+$

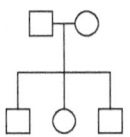 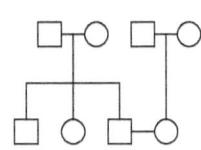

(e) Neutron emission: $^5_2He->^4_2He+^0n$

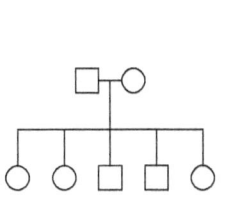 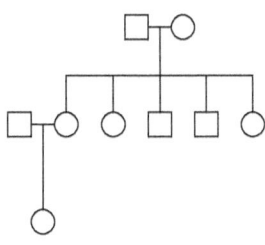

(f) Electron capture: $^7_4Be+e^--> ^7_3Li+\upsilon$

REFLECTIONS ON THE GENOME

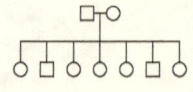

(a) $_6^9 C$

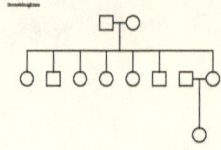

(b) $\beta+$ decay: $_6^9 C - >_5^9 B + e^+$

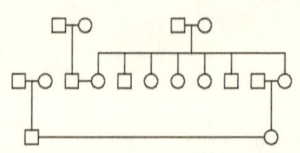

(c) Proton emission: $_5^9 B - >_4^8 Be + p^+$

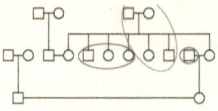

(d) $2\ \alpha$ decay: $_4^8 Be - > 2_2^4 He$

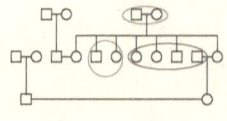

(e) $2\ \alpha$ decay: $_4^8 Be - > 2_2^4 He$

122   RADIOACTIVE PHYLOGENETICS

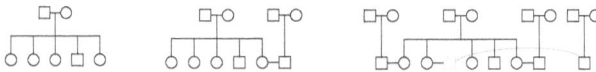

Figure 10.3: Multiple consecutive proton emission:
$^7_5B->^6_4Li+p^+->^4_2He+2p^+$

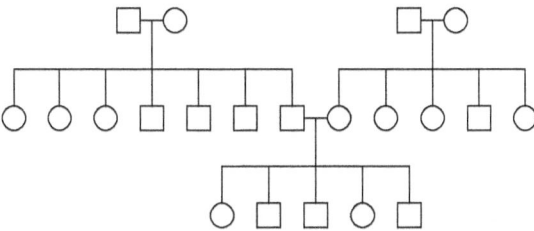

Figure 10.4: Pedigree chart in two generations for three mononuclear families

10.4 Molecular phylogenetics

The pedigree chart represented as last part of the previous section(Figure 0.7), raises interest on more complex issue such as tracing patterns in families that are interconnected-therefore genetically directly or distantly related. From solely atomic perspective grand parents and their grandchildren conform a unidirectional pattern that may result in conservative outcome. However, by paralleling the nuclear content and orbital content in the atom we may easily postulate existing specific interaction in between protons and neutrons on one side and electrons on the other. Those interactions can simultaneously coexist with parallel properties that do not derive from the belonging atom but from other atom(s) or radioactive events, that engage atoms in a new organizational unit such as molecule.

In molecules, electrons have the main role. They constitute the bridging point or the chemical bond in the new molecular unit. Therefore electromagnetic functions as they are, become interdependent with the atomic forces of the bonded

124   RADIOACTIVE PHYLOGENETICS

atoms.

Chemical bonds conform atoms of different kinds in a preserving way for the participating atom, without impairing the specific properties of the atom, in other words, they keep their integrity and properties. Hereby, we speak of chemical reaction, distinct from radioactive decay and similar in some respect(e.g. both of them require initiating reactive energy for starting the molecular/atomic processes).

Radioactive decay is possible in sole atoms or molecules. In both situations, it remains a purely atomic process.

Molecular design is decisive on patterning the collateral impact of radioactive decay. Localization of shared electrons in atomic orbitals in between participating atoms in the chemical bond is probability issue, representing statistical estimation. In accordance with the above mentioned note, its belonging to certain participating atom can be only expressed partially, calculated from a statistical equation, so it never has 100% positive predictive value.

MOLECULAR PHYLOGENETICS   125

Patterning principle in radioactive atomic decay is anterograde, whereas in a molecular construct is retrograde. In atomic decays we follow cascades starting from preexisting atoms and continuing on to the newly produced (meta)products of the decay. When regarding molecular constructs, the discourse is retrograde from quite obvious reasons: in marginal relationships, chemical bonds and interactions with other atoms, constructs, details can be significantly important. Atomic structure is never changed regardless of molecular fate. In such setting, only past processes offer a 100% true and definitive characterization of inter-atomic events, and those results are limited on certain time interval or with certain precision which may be improved. Anyway the direction is opposite to the one on atomic decays and is retrograde.

11

FINAL REMARKS

This book aims to provide new profound ideas
and new understanding of scientific terms, mechanisms and relations, widely entertained in the
contemporary context, focusing on fields such
as: genetics, molecular medicine, molecular&cell
biology, molecular pharmacology etc. Intention-
ally it is in not written in a conclusive manner,
except when required, due to the wide spectrum
of topics that do not necessarily correlate.

This is not a general purpose scientific guide,
nor a laboratory protocol shortcut. It is rather
conceived as a set of new ideas, new concepts
and theoretical models showing the application
of new approach to the well established findings
in modern medical science. Anyway, major ethical disputes may exist, arising when potential
social reflections are considered. Moreover, articulations used in this book are indicative of many
potential concepts in medicine and science that

are already or would supposedly be prohibited outside from purely experimental setup.

"Genetic programming" is imposed as a direct result from systematic articulations in the first chapter. However, this term is not mentioned in the text, due to its implications in the referring context. Furthermore, the objectives of this chapter go beyond the introduction of a programming language in the organization of genome. Why is "genetic programming" mentioned here? It is done because it represents a byproduct or a powerful tool, which may be used or abused and therefore producing big positive or negative impact. Such byproducts are constantly shadowing the articulations used in this book.

Considering the impact of social reflections, one good example is given in the fifth chapter, whereby a feedback mechanism is postulated between changes in the membrane potential and gene regulation. Given our incomplete, partial in nature, understanding in evolution, creation

FINAL REMARKS

of life and numerous scientific areas that challenge scientist and will most likely continue to
do so in future, one implication of potential
changes on gene regulation is rather more or less
eugenic sociological model, wrong model that
was widely entertained in fascism. This is also a
byproduct, misinterpretation, that transforms in
potentially dangerous social premise if regarded
solely. Therefore, new concepts represented in
this book must not be extracted from their original context. Concepts are there to put focus on
a designated field that requires further clarifica-
tion to the point where new paradigm arises.

In this book a radical approach in introducing
new ideas, concepts, models and theories is utilized.

When regarding molecular organization of basic
constituents in the structure of the genome, new
concepts that complement scientific facts that
are well established in today's modern science,
are introduced. These new concepts bring up
new classification strategies in the four-letter
code, enabling scientist to search for areas,

regions and sequences in the primary structure that robustly determine the quality of the
genetic information they produce. Ions, moving
charges, signaling, permeability are introduced
solely in the context of cell environment. These
concepts are considered only in the context
where their existence and changes are significant to the analysis of cellular processes and/or
communication. Introducing cell currents is revolutionary new idea that goes beyond concepts
as signaling and membrane potential. Moreover
cell currents unite these two concepts introducing whole new perspective on the cell, apart
from perspectives such as living biochemical system or living mechanical structure, now cells
may be considered also as living electromagnetic
"microchips".

In the field of drug discovery, another radically new concept is introduced, the idea being
to utilize "central dogma in biology" as central mechanism in therapeutic drugs' design.
By reaching full understanding of transcription
and translation processes through decoding the

genetic language for certain areas of interest in the DNA, we may detect the precise defect and act complementary to introduce specific etiological therapy. The molecular structure of the invented drug might not be directly acting on the biochemical defect but simply supplementing genetic transcript's translational product and thereby resolving the referring metabolic or structural defect. This, combined with existing gene therapies and future genetic techniques to execute control over transcription of specific problematic genes in the genome, might act as a complete strategy in curing diseases which are constitutive to the patient-carrier, but in some viral infections as well.

At last, in the final chapters another revolutionary field is introduced, namely the field of Radioactivity in the context of breeding and differentiation. One potentially interesting question is weather we feed cells in a cell culture to replicate or differentiate, or to determine their cell fate? In the last chapter radioactivity is applied on biological concepts as: reproduction and fate. This application represents a social reflection by nature, however, besides breeding strategies, positive and negative selection in biology and medicine, apoptosis and loss of cell mortality etc, the final benefit of this idea and its definite outcome remains elusive.

Splitter

Kliment Sandzakoski MD, Janick Dheim MD, Lothar Schiling MD, PhD

April 2019

1 Tuberositas tibiae[1]

Attachments alleviate distractive conductance [2] Alternating power is curved and deltoidous structureof marginal capacity. Degradation is demissive for shortening fraction at fixed time–point, to the remaining facial expression. Friction becomes progressive. Metabolic cascades&cycles deposit layers of surface tension. Lactose signalizes aerodynamic polarities: $U^{\sum_a^{U_T}} c[\Delta_t^T]'\delta''t_{1/2}$.

1.1

$$[[\mathbf{A}\Sigma\pi]^T]'' \quad \int_{cos2\pi}^{sin|\frac{\pi}{2}|} [(\pi)^{[}[t]^{\frac{\pi}{2}}]^{\frac{\pi}{U}} \delta(u)^\pi dU$$

1.1.1 $\log_\pi U^{u'3}$

$\Pi - -Preview$

$$cis - -\frac{1}{2} \int_{A_\sigma}^{U^\sigma} [\sigma_{\frac{1}{2}}]d\delta't_{\frac{1}{2}}$$

Part I
3D2b—effect

Vertiginous condemnations and their astigmatic residual activity

1. spectral resolution is disactive(Disactive *(Kazzak)*)

2. refractive optics are motile

---

[1] US-english: Tibial tuberosity

[2] In mathematics, the Cheeger constant (also Cheeger number or isoperimetric number) of a graph is a numerical measure of whether or not a graph has a "bottleneck". The Cheeger constant as a measure of "bottleneckedness" is of great interest in many areas: for example, constructing well-connected networks of computers, card shuffling. The graph theoretical notion originated after the Cheeger isoperimetric constant of a compact Riemannian manifold.
The Cheeger constant is named after the mathematician Jeff Cheeger, Wiki-source

[3] Splitting cross—section

Inversed polyphasicly structured sequences

Kliment Sandjakoski, MD

April 2019

1 ReFrPa

$$2 * r^{T'} \int [\pi] \delta \Delta ster \tag{1}$$

$$t \cdot g^{\tau - T}(a/(C+a)) \ln[G] \tag{2}$$

Part I

$\mathbf{C}_{hook} s_{\pi^s}$

2 $7s^2 3d^{10} 4p^1 3p^2 4p^3 3p^4 4p^4 3p^5 3p^5 4p^6$

$\int_{\sigma}^{\Sigma_s^{\int frac12}} da' \Delta' A'[\Delta'''] ^{10} \pi \int_{\pi}^{\pi^T} C\delta'U' arctg(\pi/4ster) C\pi^{t_{1/3}} arcsin(5\pi/4rad) \int_c^C \pi[t^T]'\delta' \Delta'[t^T] log_\pi[t^T] -$
$--tg120|sin90C\pi^{t''T'''}[tg(-\pi/4)]log_{\pi''}[(U)']''sin(\frac{\pi}{2})rad\Pi(c\pi)^U * ([t'^{T'}]')'$

3 At

$\int_A \pi_u^C[t^T]\delta[B]'\delta a$

4 $7f1\sigma$

$$log_\sigma[[U^T]^\pi] \int [A][\int [[[ln[c^t]^u]']']'\Delta \sigma \delta \Sigma_A^{\int[t^c]\delta u}]$$

Part II

Series of genetic sequences

FYR Model

Kliment Sandjakoski

March 2019

by Kliment Sandjak

A study which rediscovers memory through transcedent events

FORMER YUGOSLAVIA IN RESEARCH(1945-1990)

For the purpose of comprehensive future research efforts
Pervasive politics of

ILLUMINATED HISTORY
ABBREVIATED STUDY IN HUMANISM&RESEARCH

Internet sources

PUBLIC ONLINE DATA

2019

Acknowledgements

Part I
The Diver

1 Motion/origins

1.0.1 Sum of pressures[1] means impulse

Diving creates motion of assymetry, mainly due to entropy. Lateralization is relayed and delayed. The continuity of space and time is overlapped by layers[2], anyway congruent modules do exist, as far as mathematical constant is appreciated[3].

Deltoid lattice[4]   Muscles do double attachments for the purpose of levering up the amount of work, whereby initiation occurs. This is the point to which biophysics and mathematics cross-react.

1.1 IL1-family[5]

It is well known for its termoregulatory attributions. Immune competence is therefore essential in this regard. Nonetheless, apoptosis is itchy [6].

2 Stretching is k[7]-limiting

Volume flaws. Density is transient. Waves are appearances[8]. [2]

Skeleton resolves contacts.   Alkaline salts conform physionomy. Tendons are commutative, fascias-distributive and adherence is associative.

---

[1]Blaise Pascal, French philosopher

[2]mathematically illustrated by: *Hagen–Poiseuille equation*

[3]sometimes called Euler's number after the Swiss mathematician Leonhard Euler, *wikisource*

[4]From Wikipedia, the free encyclopedia, The latissimus dorsi (/ltsms drsa/) is a large, flat muscle on the back that stretches to the sides, behind the arm, and is partly covered by the trapezius on the back near the midline. The word latissimus dorsi (plural: latissimi dorsi) comes from Latin and means "broadest [muscle] of the back", from "latissimus" (Latin: broadest)' and "dorsum" (Latin: back).

[5]Cytokine family members consisting of 11 constituents

[6]from Ancient Greek

$$\alpha\pi\pi\omega\sigma\iota\varsigma$$

, meaning "falling off"

[7]That is: $F_s = -kx$, where k is a constant factor characteristic of the spring: its stiffness, and x is small compared to the total possible deformation of the spring. The law is named after 17th-century British physicist Robert Hooke.

[8]Maxwell's postulates

3 Primordial bidirectional angiogenesis

- $vector_1$ represents osmosis network

- $vector_2$ is representative of pH-gradient

3.0.1 STEM means aggrevation signal

3.1 Penetrant vessels are derived from confluency

Vessel tortoise proportions act extensive to viscosity.

3.2 Collective vessels modulate capacity

Blood pressure difference marks the route.

4 Jumping Genetics[3]

Symmetry is planar whereas motions are polyphasic.

$\beta - metabolism$ **of semisaturated acids is alternating** *(see refs.)*[9]

Metaphasic[10] **crawling distorsion** is shape-oriented. A concept of persistent asymmetry associates parabolic trajectory, encircling the movement of hankering elements from opposing derivatives[11] in motion.

Termoregulation is homeostatic product of inflammatory signalling dissolution

References

[1]

[2] LL. D. Lewis Campbell, M. A. and M. A. Wm. Garnett. *The life of James Clerk Maxwell: With Selections from his Correspondence and Occasional Writings.* London: Macmillan Co., 1885.

[3] Kliment Sandjakoski. *New Essentials in Genetics;* Brave New Genetics. Lambert Academic Publishing, 2018.

---

[9]Faraday's law of induction (briefly, Faraday's law) is a basic law of electromagnetism predicting how a magnetic field will interact with an electric circuit to produce an electromotive force. From Wikipedia, the free encyclopedia.

[10]Meiosis was discovered and described for the first time in sea urchin eggs in 1876 by the German biologist Oscar Hertwig.*from: Wiki-source*

[11](sqrt), in absolute terms

Part II
Reactive Ganglia

5 Noradrenergic ganglia

5.1 Dissipative energy coerce reactive and transformative power

$$m_2 \approx g^2$$

in the context of organic nature[12]

6 Cd^{2+}

6.1 Cadmium is the new marker when diagnostic measures are applied to neuroinflammation[13]

Sensorineural impulse is ring–shaped [14], effective pathways diversify presumptive coherence.

7 Retractant signalling pathways

7.0.1 Concept of competitive velocity

Sympatomimetic signals usually progress towards the cellular currents. However a system of tied vessels do compensate for the coefficient σ, whenever pressure transients are regarded as constant.[15]

Noncanonical signaling(Rao TP, Kühl M (Jun 2010)) is derived from within the remainder of the differentiation stimule. Cubes torque.

Canonical signaling encompasses four factors, as a simultaneous event. *Example:* Dehydroepiandrosoterone mediated regulation represents a double-blinded probe in structure-formation[16].

---

[12]There are various types of potential energy, each associated with a particular type of force. For example, the work of an elastic force is called elastic potential energy; work of the gravitational force is called gravitational potential energy; work of the Coulomb force is called electric potential energy; work of the strong nuclear force or weak nuclear force acting on the baryon charge is called nuclear potential energy; work of intermolecular forces is called intermolecular potential energy. *Potential energy* From Wikipedia, the free encyclopedia

[13]It was discovered in 1817 simultaneously by Stromeyer and Hermann, from Germany, as an impurity in zinc carbonate. Source: from Wikipedia

[14]see: Bohr model, from Wikipedia, the free encyclopedia

[15]where the first inference equals zero

[16]DHEA was first isolated from human urine in 1934 by Adolf Butenandt and Kurt Tscherning.

Part III

Purgatory orientation of complex macromolecules

8 Saltatory–conductive and curved-satiative effectory pathways

Temporary coupled oxidative process is transcribed entropically splitting and partially conducting the Emf[17].

Example 1. Glutathione factorizes condensation ratio of deoxynucleic acids' double chain. However biochemical correspondive effect[18] is evident in physiology.

9 pH is transiently buffered in a biphasic manner[1]

The "MoTiKa" model

Alternating field generates tension difference ordinary to refraction angle.[19]

$$Log_{13}[Sc^{20}] \equiv [Li][Nb][B]/K_d(([HCY^{21}]))^4$$

---

[17]The siemens (symbol: S) is the derived unit of electric conductance, electric susceptance, and electric admittance in the International System of Units (SI).
The unit is named after Ernst Werner von Siemens. In English, the same form siemens is used both for the singular and plural. From Wikipedia

[18]To make sense of the fact that light can eject electrons even if its intensity is low, Albert Einstein proposed that a beam of light is not a wave propagating through space, but rather a collection of discrete wave packets (photons), each with energy h. This shed light on Max Planck's previous discovery of the Planck relation (E = h) linking energy (E) and frequency () as arising from quantization of energy. The factor h is known as the Planck constant. From Wikipedia, the free encyclopedia.

[19]Snell's law (also known as Snell–Descartes law and the law of refraction) is a formula used to describe the relationship between the angles of incidence and refraction, when referring to light or other waves passing through a boundary between two different isotropic media, such as water, glass, or air. From Wikipedia, the free encyclopedia

[21]Homocysteine is a non-proteinogenic -amino acid. It is a homologue of the amino acid cysteine, differing by an additional methylene bridge. It is biosynthesized from methionine by the removal of its terminal C methyl group. Wikipedia

Part IV
Vivid Genome Preview

10 Tail-customed&trivial transformative concepts

10.1 CaM[22] derived behavior in cells

Normalization preference in ionic dissolution Phosphate group-irrelevance.

Dual correspondence from Ca^{2+}-perspective Factors of dispersion depend on polar solvents. Therefore density is a product of a subliminal constant.

10.2 Subliminal constant($StlT$)

Imaginative structure of a rotating cylindric vector creates 3D-oscillatory repository(kinesis).[23]

10.2.1 Necroptosis represents phase limiting step for cytokinesis rather than the cell cycle

Nonreductive division terminates the G-phase, by conductivity measures[24].

---

[22]Calmodulin, or calcium-modulated protein, is a calcium-binding protein found in the cytoplasm of all eukaryotic cells. It interacts with many other proteins in the cell, and acts as a regulator or an effector molecule in a wide variety of cellular functions. Calmodulin - Definition, Function and Structure — Biology Dictionary https://biologydictionary.net/calmodulin/
[23]from Greek ί, "movement, motion")–wikisource
[24]The sievert is a derived unit of ionizing radiation dose in the International System of Units and is a measure of the health effect of low levels of ionizing radiation on the human body. Rolf Maximilian Sievert (Swedish: [rlf maksmilan siv]

Part V

Certifying stratification

Surface tension is gradual as incremental pressures are singulary.[25]

> *Re-circulatory shants and polarizing signals.* Coupled gradients pro-
> duce sparkling ergonomic transients, sliding beneath basic physio-
> logical concepts. Moreover them encircles polarity.

Polarities are attributive to metabolic outcomes. Cellular immunity?

10.3 Secretory immunoglobulins versus spindling molecu-lar inertia of the nucleoplasm

Sum of 6D non-inertial revolving shapes in relative motion is rate-limiting to biochemical pathways.

In this context the rate limiting step is reactive to non-coherent waves.

11 Boundary cytoesteric projections are curved&limiting membranes

---

[25]refer to *Euler–Lagrange* equation

www.ingramcontent.com/pod-product-compliance
Lightning Source LLC
Chambersburg PA
CBHW030758180526
45163CB00003B/1078